The Pillars of Creation

Giant Molecular Clouds, Star Formation, and Cosmic Recycling

Martin Beech

The Pillars of Creation

Giant Molecular Clouds, Star Formation, and Cosmic Recycling

 Springer

Published in association with
Praxis Publishing
Chichester, UK

Martin Beech
Campion College
At The University of Regina
Regina, SK, Canada

SPRINGER-PRAXIS BOOKS IN SPACE EXPLORATION

Springer Praxis Books
ISBN 978-3-319-48774-8   ISBN 978-3-319-48775-5   (eBook)
DOI 10.1007/978-3-319-48775-5

Library of Congress Control Number: 2016962718

Cover design: Jim Wilkie

Printed on acid-free paper

This Springer imprint is published by Springer Nature
The registered company is Springer International Publishing AG
The registered company address is: Gewerbestrasse 11, 6330 Cham, Switzerland

Preface

The human spirit is the lamp of God,
Searching all the innermost parts.

—Proverbs 20:27

Human knowledge grows and science works by casting its proverbial gaze ever forward, searching out those dark recesses, holding the lamp of understanding as high as possible, and casting its light as far as it will go, looking for the new and novel. It is at those quick-silver moments of new illumination that we arrive at the very hinge of discovery—caught in the ever-moving now, sandwiched between the weight of the past and the untouched future—we push our way forward, edging toward the previously unseen.

By casting our gaze toward the deepest shadows, the shadows that lurk within the realm of the unknown, we also illuminate, just a little more, the trail of history. The context of the past becomes ever more clear each time we move forward, and the discovery of the new behooves us to look back and ask pointed questions, such as how did we get here? Sometimes our new view of the past reveals that a crooked, roundabout, road to discovery has been taken, and sometimes it illuminates a better path—an alternative highway not shown on the map we originally had in hand—a road, as it were, drawn in invisible ink, requiring just the right illumination and viewing conditions for it to be seen. Sometimes, by casting light upon the unknown, we illuminate a cul-de-sac (a literal Tolkienian *bag end*), and it is real-ized that a wrong turn has been made and we can go no further. At these moments, there is no choice but to start again, retrace our steps, and start over, the trail starting out in a differ-ent direction. The new road may take us further than before; it may not.

This is how science works: ask questions, collect data, and take mental journeys of mathematical and cognitive adventure. Some questions of the world around us are easy to formulate and profound in their depth but incredibly difficult to answer. We will likely never know, for certain, how the universe came into existence, but we can assuredly ask detailed questions and expect reasonably detailed answers, of the phenomenon and objects that we presently see within the fold of the cosmos. In the story to follow, we take it as fact that the universe had a moment of specific creation, bringing into existence, about 13.7 billion years ago, a vast reservoir of raw materials. We ask not how this material came into existence but, rather, what has happened to all of that primordial material since it was first created. In short, we are asking, what is the history of star formation?

To answer this seemingly straightforward question, we will need to take a number of long journeys, both in time and space. The narrative to be developed will require the bringing together of material from many notebooks, diaries, textbooks, and journals—the tactile repositories of scientific journeys of exploration. Most of all, however, our narrative will be built upon the derived knowledge of the stars, the interstellar medium, and giant molecular clouds. The primordial nucleosynthesis that followed the first moments of creation may well have produced the basic building materials, but it is the circumstance of star formation that has allowed the universe to shine and sparkle with possibility. Stars, as we shall see, are the great sub-creators. They make chemistry possible, and it is them that have enabled the conditions for planets and life to come about. They are the veritable engines of creation. Although the Big Bang provided the fecund spark of initiation, the primordial universe that it sired was born hopelessly sterile, and it is only by the continued recycling of the interstellar medium, through star formation and stellar evolution, that the universe has been animated beyond a chaotic mess of elementary atomic particles, radiation, dark matter, dark energy, and expanding spacetime.

Physical measure and human ingenuity have, over the eons, separated out the stars from the planets, but throughout most of human history, cradled within a spherical shell encircling the Sun just beyond the sphere of Saturn, the stars were close and familiar. It is hardly surprising that our distant ancestors believed that the stars could influence the very ebb and flow of life. Indeed, by modern standards, the stars, for them, were almost touchable. The wonderfully complete picture of the classic medieval universe has now long been discarded. It no longer fits the facts as revealed by numbers and data, the indisputable truth of measure having overruled human sensibility, suspicion, and delusion. The stars are distant, the stars are remote, and they care not for humanity. What we see in the heavens today is the very same matter that was seen by our distant ancestors, but now, we see deeper and we know much more about what it is that we observe. In a sense of reverse astrology, the stars no longer dictate our future, but it is us that know and tell of the full cycle of their lives. How mightily the times have changed.

Figure P.1 shows one of the most remarkable astronomical images ever taken. It is the Hubble Deep Field. Almost alarmingly, given the rapid pace of modern scientific research, this image, which was released by NASA on January 15, 1996, is old hat. There are now the Hubble Ultra-Deep Field and Hubble eXtreme Deep Field[1] images. The Hubble Deep Field image, however, was the pioneer. It was the first image to catch the collective public imagination, literally forcing the viewer to contemplate their infinite smallness in contradistinction to the incredible vastness of the observable universe. Humanity is humbled by

---

[1] Although the images and data evolve rapidly, it is almost comforting to note that NASA and research scientists in general continue to mutilate the English language and generate spellings and acronyms through the seemingly random sampling of words and letters.

Fig. P.1 The Hubble Deep Field. This image was assembled from 342 separate exposures taken over 10 consecutive days starting December 18, 1995. The field of view covers an area of about 7 square arc minutes on the sky and corresponds to about one 24-millionth of the entire backdrop of the heavens. (Image courtesy of NASA/ESA/HST)

images such as the Hubble Deep Field, although our collective id happily survives the implications.

Taking the incredible image that the Deep Field is and converting it into the equally beautiful dataset of numbers that it contains, it is revealed [1] that there are less than 20 individual stars in the image but some 3000 images of galaxies. The most distant galaxies are located over 12 billion light-years away, and a detailed analysis of galactic distances and light emission reveals that star formation was much more rampant in the distant past, attaining a peak some about 8–10 billion years ago. By the time the Sun was formed, some 4.5 billion years ago, star formation in the universe was on the wane, which is not to say that star formation is going to end any time soon, as we shall see later on.

Coming back to the raw numbers, the Hubble Deep Field enables us to make an estimate of the number of protons and electrons that exist in the observable universe [2]. This is an incredible result, since these particles are the basic building blocks of all matter, and they constitute the matter that we, as human beings, can physically experience, whether by feel or sight. Remarkably, the number of protons and electrons in the observable universe can be estimated in a space no larger than the back of an envelope (or, for the more modern readers, a space the size of a cell phone screen). The calculation is actually quite straight-forward (the very hard bit is actually getting the observational data). The area covered by the Deep Field is about 1.9×10^{-3} square degrees on the sky (which is about the same area as that covered by the following wingdings square □ seen at a distance of 4 m [3]), and the mathematicians tell us that there is 41,253 square degrees over the entire surface of a sphere. Taking the Deep Field to be a typical view of the observable universe, this suggests that the number of galaxies N_{gal} in the observable universe is $N_{gal} = (41,253/1.9 \times 10^{-3}) \times 3000 = 6.5 \times 10^{10}$, with the 3000 coming from the actual galaxy count in the Deep Field. This number tells us that our Milky Way galaxy is just one of at least 65 billion galaxies in the observable universe.

Let us suppose, and these numbers will be justified later, that each galaxy contains 100 billion stars ($N^* = 10^{11}$) and that the typical mass of a star is half that of the Sun ($M^* = 10^{30}$ kg). The physicists tell us that the mass of the proton is $m_p = 1.7 \times 10^{-27}$ kg, so a typical star (assumed to be made entirely of hydrogen) will contain of order $N_P = M^*/m_p = 6 \times 10^{56}$ protons. We can now determine the approximate number of raw matter building blocks—protons and electrons—that there are in the universe: $N_{pe} = 2 \times N_{gal} \times N^* \times N_P = 8 \times 10^{78}$; the 2 in the calculation accounts for the fact that there is an equal number of electrons and protons;—the universe has no net electrical charge.

As one would expect, the number of protons and electrons is staggeringly large, but equally as staggering is the fact that we can estimate their number (to within a factor of about 10). For our story, however, it is the long-term fate of these 8×10^{78} protons and electrons that is of interest [4]. These myriad miniscule bodies are the building blocks of the stars, and the question to be explored in the following chapters is how such groupings of 10^{57} protons and electrons are formed, transformed, and recycled.

The dome of the celestial sphere is never more exhilarating than on a clear winter's night. The stars are bright and crisp on the sky, and it would seem that their number is uncountable. Dominant among the wintertime constellations is Orion, with his three star-studded belt, long legs, and outstretched arms. We have all seen Orion and marveled. And, indeed, when we look at this gangly limbed constellation, we are catching the light from stars in the making, and lurking in the background, hidden to the gaze of our eye, is a massive molecular cloud that covers the entire vista.

As you read this page, stars are being formed in the great Orion Nebula, located (as we see it) at the tip of the hunter's sword extending downward from the middle star on his belt. The light from these sibling stars, however, is already old by the time we get to see it. Indeed, located about 412 pc from the Sun, it takes over 1300 years for the new starlight to reach us. The light we see now left the nebula in our eighth century, at a time when the heroic poem of *Beowulf* was first being chanted by flickering firelight, the classical Mayan civilization was at its peak, the Venerable Bede was to complete his *Ecclesiastical History of the English People*, and Charlemagne was to be crowned the first Holy Roman Emperor. Much has happened on Earth since the light we now see from the newly formed stars in Orion started journeying outward.

Although the light from the Orion Nebula has been visible since humans first gazed toward the winter night sky, it was only recognized as being a cloud of glowing hot gas in 1864, when William Huggins, working from his Upper Tulse Hill, London, observatory, was able to resolve emission lines within its spectrum. This important discovery heralded the identification of the interstellar medium, and subsequent study by many generations of astronomers has revealed that our Milky Way galaxy is parturient with star-making gas and dust. Indeed, the disk of our galaxy, like an exuberant Rubenesque goddess, exudes great dimples and folds of interstellar matter, and it is the remarkable cyclical journey of this material, from gas cloud to star and back to gas cloud again, that this book will explore.

The Sun, the nearest star, is fundamental to life on Earth and humanity. It has nurtured our long chain of being, and it has, at least thus far, supported our faltering steps toward transcendence. Earth is our home, and our backyard is the Solar System. We currently live on the former, while the latter offers the opportunity of resources and possibly future domicile. The Sun dominates the Solar System; it is where virtually all the mass resides. Just 0.1 % of the Solar System's mass is tied up in the planets, comets, asteroids, and Kuiper Belt objects; the other 99.9 % is in the hot and roiling body that is our Sun.

However, the Sun is neither a typical star nor is it entirely located at random within the Milky Way galaxy. It is assuredly a special star. Nature mostly makes stars that are not like the Sun. Indeed, upon the galactic scale, the Sun is just one of hundreds of billions of stars, a rather massive star (but far from being the most massive star possible) but also a relative newcomer, being a mere 4.5 billion years old—a timescale that spans about the last one-third of the Milky Way's current age.

It is now well known, and readily stated, that without the Sun, there would be no life on Earth. But the story runs much deeper; without long-dead stars, there would be no hope for life anywhere in the universe, and without galaxies, those vast assemblages of stars that pepper the vast volume of the cosmos, there would be no planets and no ongoing star formation. Indeed, it is the self-gravity of our Milky Way galaxy that keeps the raw materials for the making of stars and planets, the interstellar medium, in place. Our existence depends not only upon the small things, our DNA, the rush of chemistry, and the slow-searching fingers of evolutionary adaptation. But it also depends upon the largest structures that we know of in our galaxy's disk, the giant molecular clouds and the stars that form within their glowing nurseries.

This book is about the unfolding story of star formation and how stars are intimately linked to the interstellar medium out of which they form and eventually die—literally as ashes to ashes and dust to dust. For certain, we do not know the full and detailed narrative of the star formation process, and we certainly have not, as yet, asked all the right questions of the phenomena so far discovered. But the journey of discovery continues apace; we are speeding onward, casting an ever-widening circle of exploration and understanding, and the vistas that are unfolding before our gaze are truly humbling and excitingly spectacular.

Regina, SK, Canada   Martin Beech

REFERENCES

[1] See, for example, H. C. Ferguson's detailed article, The Hubble Deep Field, published in *Reviews in Modern Astronomy*, **11**, 83-115 (1998).

[2] This estimate accounts for the so-called baryonic matter – that is, matter made of three quarks – contained within stars. Observations indicate that at least five to ten times more matter again, is contained within the hot, diffuse intergalactic medium. The term observable universe is used here since this corresponds to what we can actually see – the universe may of course be much larger than we suppose. The topic of dark matter and the effects of dark energy have been discussed by the author in, *The Large Hadron Collider – Unraveling the Mysteries of* Universe (Springer, New York, 2010).

[3] In 10-point font size the *wingdings* square is about 3 millimeters on each side.

[4] This number is often called the Eddington number, after Arthur Eddington's famous calculation for the number of protons and electrons in the universe. Eddington, however, derived his value in an entirely different fashion, not actually based upon star or galaxy count data. The rather obscure reasoning behind Eddington's calculation is explained in chapter 11 of his posthumous text, *The Philosophy of Physical Science* (Cambridge University Press, 1939). Eddington argued that the number of protons in the universe was given according to the number 136×2^{256} – an 80 digit number that he evaluated by long-hand during a steamship crossing of the Atlantic.

Contents

1

Reading the Sky

There, in the night, where none can spy,
All in my hunter's camp I lie,
And play at books that I have read
Till it is time to go to bed

—R. L. Stevenson

Invoking Buridan's Donkey

The year is 1692, and the reverend Richard Bentley, Master of Trinity College, Cambridge, has just asked Isaac Newton a deep and troubling question. An acclaimed theologian, Bentley was preparing to deliver the first of the newly endowed Robert Boyle Lectures. Open to the public, this series of eight lectures was to be delivered from the pulpit of St. Martin's Church, London, and the talks were to consider the role of Christian theology and the requirement of an active deity within the framework of natural philosophy (what we now call science) and specifically within the quest to understand the workings of the universe.

Bentley's last three lectures struck to the core of his correspondence with Newton. By 1692 Newton's magnum opus, the *Principia*, had been in print for only 5 years, and while Newton's ideas and explanation of gravitational attraction were still new and hardly known beyond a small circle of the scientific elite, Bentley was interested in how this all-pervasive force might influence the stability of the universe. Specifically, Bentley asked that if gravity acts universally, and if stars are corporal bodies, then why are the heavens not alive with motion, with the stars streaming in every which direction? It was (and is) a good question, and Newton could offer only a vague and contrived answer. The reason why the stars appear static, Newton reasoned, was because the universe is infinite in extent and because the stars are uniformly arranged within its interior—the stars being set, to use a modern perspective, something like the atoms in a massive, continuously repeating crystal lattice. In short, Newton was suggesting that although every star did indeed feel the gravitational tugs from all of its nearest neighbors, the resultant pull, because of the uniform distribution, was equal in all directions and effectively summed to zero. The stars, just like Buridan's philosophical (and hard done by) donkey, feeling no compulsion to move one way or another, simply stay put.

© Springer International Publishing AG 2017
M. Beech, *The Pillars of Creation*, Springer Praxis Books, DOI 10.1007/978-3-319-48775-5_1

Newton realized, however, that this arrangement must be highly unstable: it was a veritable house of cards. If just one star anywhere in the infinite depths of space shifted only slightly in its position, then the consequences would be fatal, and soon all the stars would be in chaotic motion. The perceived constancy, therefore, of the heavens and the apparent stability of the stellar system, from Bentley's and indeed Newton's, approved viewpoint, was a clear indication that an active, intelligent, and omnipotent Creator must be at play in the cosmos. Theology and the new physics of Newton combined, in the case of Bentley's Boyle lectures, to argue for a cosmos that was composed of a static and uniformly spaced infinite system of stars.

Halley's Lucid Spots

The first two decades of the eighteenth century were a remarkably productive time in the remarkable productive life of Edmund Halley. Halley turned a sprightly 44 years old as the new century began, having spent the last several years at sea as captain of HMS *Paramour*—a 6-gun pink of the Royal Navy. During his command of the *Paramour* Halley had sailed (not always with a harmonious crew) the South Atlantic seas, measuring variations in Earth's magnetic field [1].

Upon retiring his naval commission in 1703, Halley was elected to a professorship of geometry at Oxford University, and within short order he produced his *Synopsis of the Astronomy of Comets* (1705), learned Arabic and finished a translation of Apollonius's *Conics*, and outlined (1716) a method by which the size of Earth's orbit might be determined by viewing transits of Venus. In 1718 he made the monumental discovery of stellar proper motion—showing that stars[1] do indeed move relative to each other in the sky, and that they must also be moving through space. In 1720, along with antiquarian friend William Stukeley, Halley assisted in the first scientific attempt to determine a construction age for the great monoliths of Stonehenge, arguably one of the earliest custom-built astronomical observatories.[2] It was also in 1720, following the death of the cantankerous John Flamsteed, that Edmund Halley accepted the post of Astronomer Royal at Greenwich Observatory in London.

In between all these monumental works and activities, Halley also found time in 1715 to present an account to the Fellows of the Royal Society on "Several Nebulae or lucid Spots like Clouds, lately discovered among the Fixt Stars by help of the Telescope." Although arguably one of Halley's lesser papers—just three pages in extent when published in the *Philosophical Transactions*—his account drew attention to some of the new novelties that were then being discovered in the night sky. Halley writes, "[these] wonderful… luminous Spots and Patches… are nothing else but the Light coming from an extraordinary great Space of the Ether; through which a lucid *Medium* is diffused, that shines with its own proper Lustre."

---

[1] Halley specifically found that the positions of the bright stars Sirius, Arcturus, and Aldebaran had moved over half a degree on the sky since the compilation of the star catalog by Hipparchus circa 135 B.C.

[2] There is little doubt that Stonehenge was primarily constructed to accommodate the religious and funerary practices of a specific Stone Age society. It was not an astronomical observatory, although its geometry does indicate alignments with the rising and setting locations of the solstice Sun.

In total Halley's catalog of nebulae contains just six examples, of which he notes that there are undoubtedly more "not yet cometh to our Knowledge." The list of objects, while small, contains, in modern terms, a star formation region (or an HII nebula), a galaxy, three globular clusters, and one open cluster. Halley writes, "The first and most considerable is that in the Middle of *Orion's* Sword… [composed of] two very contiguous Stars environed with a very large transparent bright spot," and this, of course, is clearly the now famous Orion Nebula, first revealed as being a non-star-like object through the telescope by Christiaan Huygens in 1656 (Fig. 1.1a).

Third in Halley's list of objects is the Andromeda Galaxy (technically, at a distance of 2.54 million light years away, the most distant object visible to the unaided eye), described (Fig. 1.1b) as having, "no sign of a Star in it, but appears like a pale Cloud." Fourth in Halley's list is the globular cluster ω Centauri (Fig. 1.1c)—discovered, in fact, by Halley in 1677 while mapping the southern stars from the island of Saint Helena.

Fig. 1.1 (**a**) The Orion star formation nebula. (**b**) The Andromeda Galaxy—a spiral nebula, (**c**) ω Centauri globular cluster. (**d**) The Wild Duck cluster. Images a, b and d courtesy of NASA. Image c courtesy of ESO

The fifth entry in Halley's list of nebulous objects is known to present-day astronomers as the Wild Duck cluster (in the constellation Scutum), and it is an example of a rich, galactic cluster of stars (Fig. 1.1d).

To Halley, and to all of his contemporaries, the clouds and spots identified in his 1715 catalog (Fig. 1.1a–d) were objects of curiosity, and while they were clearly not stars, they nonetheless moved within the stellar realm. Halley notes, "[S]ince they have no Annual Parallax, they cannot fail to occupy Spaces immensely great, and perhaps not less than our Whole Solar System." Halley was certainly correct in his surmise about the vast size of nebulae regions, but he was woefully off with respect to the true scale; the nebulae occupy regions of space vastly larger than our Solar System.

In his short catalog of nebulae Halley makes no attempt to relate them to any specific stage in the life of a star. Indeed, Halley knew that new stars (the supernovae of 1572 and 1604) did appear from time to time, and he also knew that some stars (e.g., omicron Ceti, or Mira, Algol and χ Cygni) were variable in their brightness. But to the question of star formation and star death he, like his contemporaries, remained steadfastly silent. Importantly, however, by the early to mid-eighteenth century it is apparent that astronomers were becoming increasingly aware of oddities in the sky: luminous clouds, new stars, and stars that showed periodic changes in brightness. What such observations and objects implied about the structure of the cosmos was then unclear, but the notion that the stars were immutable, ageless objects, and that the spaces between the stars was entirely empty, was being brought into question.

Although silent on the issue of star origins, Halley did question Newton's 1692 proclamation that the universe was infinite in extent and contained an infinite number of stars. His argument was "rather of a Metaphysical than Physical Nature," and revolved around the use of the word "infinite," Halley essentially noting that to produce an infinite number of stars in an infinitely large universe would presumably require an infinite amount of time, and this ran counter to the then accepted wisdom. Accordingly he asserted that the number of stars must be finite. Interestingly, as well, Halley ponders the question, but provides no answers or conclusions concerning the psychological consequences of residing in a region of space illuminated by a cloud of "lucid medium." He notes in particular that, "In all these so vast Spaces it should seem that there is a perpetual uninterrupted Day, which may furnish Matter of Speculation, as well to the curious Naturalist as to the Astronomer." Indeed, the question is not an entirely idle one, and we shall come back to consider the consequences of living in a dense molecular cloud region at a later point.

Wright's *Via Lactea*

A broad and ample road whose dust is gold,
And pavement stars, as starts to thee appear,
Seen in the Galaxy, that milky way,
Which nightly, as a circling zone thou seest,
Powder'd with stars

—John Milton, *Paradise Lost* (Book VII)

The picture of an infinitely repeating pattern of equally spaced stars within an infinitely large universe is not an unholy construct (and is no more bizarre than some of the more rococo cosmological models that have been presented in recent years). The minds-eye view is simple enough, easy to visualize, and it certainly has the feel of a well constructed arrangement. The problem, of course, is that the picture is entirely wrong. Newton and Bentley erred in their impeccable, though rather contrived logic.

From a modern perspective (with the benefit of several hundred years' worth of hindsight at our disposal), we can raise any number of objections to the proposed cosmological scheme developed by Newton. Firstly, it assumes that all stars must be of the same mass, and that the star formation process, even if regressed to sometime in the near infinite past, must have been extremely rapid, the stars congealing at their lattice points so rapidly that they did not have time to move before their surrounding closest neighbors formed to enforce equilibrium. Still the main problem, of course, is that the model as suggested simply looks nothing like the star distribution we actually see on the sky. The stars are not uniformly spaced from each other, and they seem to be dotted at random across the celestial sphere. Even if we use brightness as a proxy for distance, the fainter (and therefore further away [2]) stars do not spread into an all-enveloping all-sky glow.[3]

Even though we do not see any obvious order within the distribution of stars in the sky, there are nonetheless features that betray structure. Indeed, the one very obvious feature that stands out above all others on the night sky is the Milky Way (Fig. 1.2).

Long known to ancient astronomers, the *Via Lactea* (Milky Way) appears as an irregular, faintly luminescent band, marked with dark and tendril-like blotches, that stretches around the entire celestial sphere. Less easy to see for modern city-dwelling observers but very obvious to anyone under remote, clear-sky conditions, the key question to ask about the Milky Way is, is it a band of luminescent matter, or is it made up of myriad faint stars? To the ancients the answer was unknown, and indeed, for them, unknowable, but with the introduction of the telescope in the early seventeenth century, it began to become increasingly clear that the Milky Way, far from being a strange collection of luminous matter, was really the accumulated light from a multitude of faint and remote stars seen through a trelliswork of obscuring mater.

March 12, 1610, is one of those days in which the human perception of the cosmos changed. It is the day that a then relatively unknown mathematician in Padua signed off on a small booklet given the deliberately aggrandized title *Siderius Nuncius*. The mathematician was Galileo Galilei, and the booklet, *The Sidereal Message*,[4] was to change the very practice of astronomy.

---

[3] This problem is usually described under the guise of Olber's Paradox [3]. And, this paradox is only resolved under the constraint that we live in a universe of finite age.

[4] It is clear from Galileo's correspondence of the time that he intended the Latin word *nuncius* to mean *message* as opposed to *messenger*, and indeed, it is via the former term that Galileo refers to his work when describing it in his native Italian. The *tradition* of referring to the booklet as the *Sidereal Messenger* was apparently started by Johannes Kepler in his *Dissertatio cum nuncio sidereo* (published April 19, 1610).

Fig. 1.2 The arch of the Milky Way is revealed in this half-sphere image of the sky. Towards the center of the image are the gas and dust clouds in the direction of Sagittarius. Antares, the brightest star in Scorpius, is located just above and slightly to the *left* of the image center. Alpha and beta Centauri shine prominently about two-thirds of the way along the band of the Milky Way to the *right* of the image. Just to the *right* of these two stars in the image is the dark macula of the Coalsack Nebula, which is nestled within the lower corner region of Acrus, the Southern Cross. The two luminous clouds to the *lower right* in the image are the Small and Large Magellanic Clouds. Image courtesy of NOAA Science on a Sphere

Word of Galileo's work spread like wildfire. Within 2 weeks, news of its publication had reached southern Germany, and within 5 weeks it had reached the shores of remote England—a distance of some 1100 km as the crow flies. Indeed, by 1615 *Siderius Nuncius* was available in translation format in China.

Within the pages of the *Sidereal Message*, Galileo somewhat hastily describes the observations he had made of the heavens with his newly constructed telescope. The iconic instrument of astronomy, the telescope, had dramatically appeared, seemingly out of nowhere and essentially fully developed, in northern European markets in late 1608. Galileo heard of the new device in the summer of 1609, and having rapidly developed his lens grinding skills he turned his new device, with a magnification of about 20 times, to the night sky. What he found revolutionized astronomy, and it also made Galileo famous.

In relation to our narrative, the first important result that Galileo presents is given in his introduction where he notes that, "[I]t seems of no small importance to have put an end to the debate about the Galaxy or Milky Way and to have made manifest its essence to the senses as well as the intellect, and it will be pleasing and most glorious to demonstrate clearly that the substance of those stars called nebulous up to now by all astronomers is very different from what has hitherto been thought." Galileo later explains that with the aid of the telescope the Milky Way, "that for so many generations [has] vexed philosophers," is "nothing else than a congeries of innumerable stars distributed in clusters." Galileo further presents maps of the fainter stars that he has resolved in the Pleiades, Orion and Praesepe. Perhaps surprisingly Galileo makes no mention of the nebular region in Orion, nor does he mention the Andromeda nebula (which is distinctly visible as being nebulous to the naked eye).

Although Galileo rushed his *Siderius Nuncius* into print, no doubt with the intent of cementing his priority on his new discoveries, he made no attempt to speculate upon what he had actually observed. His account was matter of fact, and while he later used his observations to challenge accepted orthodoxy, the primary role of the work was to reveal the newly observed phenomena.

Johann Kepler, on the other hand, as was indeed his way, made every attempt to speculate on the greater meaning of Galileo's new discoveries. Within 2 weeks of receiving a copy of Galileo's *Siderius Nuncious*, Kepler had penned a lengthy letter (published a few months later in pamphlet form) of reply. Entitled "Conversation with the Sidereal Messenger," Kepler set down his thoughts on Galileo's work, and though his letter oozes with praise and congratulations, Kepler also brings Galileo to task on many points. With respect to Galileo's observations on the Milky Way, Kepler writes, "You have conferred a blessing on astronomers and physicists by revealing the true character of the Milky Way. You have upheld those writers who long ago reached the same conclusion as you: they are nothing but a mass of stars." This is a gentle correction of Galileo's claim for complete novelty concerning his ideas about the structure of the Milky Way, and it is by no means Kepler's only correction and/or objection to Galileo's matter of fact writing. A copy of Kepler's "Conversation" was handed to the official Tuscan couriers on April 19, 1610, and Galileo is known to have received and read the letter, but he never penned a reply.

Itinerant lecturer Thomas Wright of Durham was one of the first astronomers to attempt an interpretation of what the Milky Way was in terms of a vast accumulation of stars. Writing in "An original theory or new hypothesis of the universe," published in 1750, Wright reasoned that the appearance of the Milky Way indicated that the Solar System must be located within a flattened, disk-like distribution of stars. The system was disk-like since the Milky Way stretched around the entire sky, and flattened since it was a localized band of concentrated starlight. Indeed, Wright draws a visual analogy with the planet Saturn and its extensive ring system. At play was an "optical effect." When an observer looked into and along the disk, many stars would be seen, but when looking at right angles, and therefore out of the disk, a relative dearth of stars would be evident.

Wright actually vacillated in his ideas about the potential distribution of stars, suggesting that the same optical effect and appearance of the Milky Way could be achieved if the Solar System was located within a vast, thin shell of stars (Fig. 1.3). Wright came up with two very different models to describe the shape of the stellar universe partly because he insisted that at the center of the universe was (naturally) the "Abode of God."

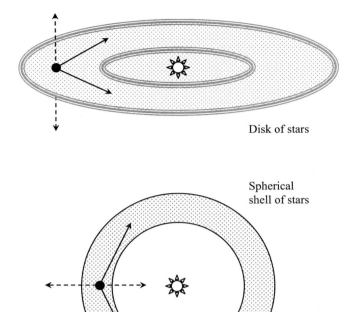

Disk of stars

Spherical
shell of stars

Fig. 1.3 Thomas Wright of Durham's two possible interpretations for the stellar configuration necessary to reveal the Milky Way. In the direction of the *dashed-line arrows* an observer
would see many stars systematically arranged in a flattened distribution on the sky, while in
the direction of the *dashed-line arrows* an observer would see few, and randomly distributed,
stars. The Abode of God, which is remote and unknowable, is located, according to Wright,
at the center of each construction

Although by present-day standards Wright's invocation is not a required pre-requisite
for a cosmological model, it nonetheless required Wright to consider the consequences of
Newton's universal gravitational attraction. Indeed, Wright adopted the stance that all
stars must be in motion. This was not an entirely new idea, and indeed, Edmund Halley, in
1718, had already shown that at least a few stars had distinct and measurable proper
motions on the sky [4]. Again, though Wright's ideas have a decidedly modern ring to
them, his reasoning is suspect, and stellar motion was invoked so that the stars could never
fall to the core of the universe and thereby impinge upon the "Devine Centre." Were the
stars arranged in some motionless configuration, Wright argued, invoking the idea first
raised by Richard Bentley some 58 years earlier, they would simple collapse under their
mutual gravitational attraction to a common center. For Wright, the stars were put in
motion about the Devine Center at the time of the Creation, and this implies that the stars

were all formed at the same time.[5] Wright's ideas are decidedly idiosyncratic, but they at least initiate the important cognitive step that takes us towards the idea of a dynamical universe (or galaxy as we would call it) with the stars preferentially arranged according to an identifiable structure.

Almost immediately after publishing his "An original theory," Wright began to backtrack on his ideas, and he eventually sold out to classical ideas, suggesting that the, "starry firmament might well prove to be no other than a solid orb … And the fixed stars no more than perpetual lamination or vast eruptions." History has mostly relegated, if it appears at all, Wright's ideas to the foot- or end-note sections of modern cosmology texts. It sits on the uncomfortable divide between what we now recognize as being real scientific enquiry (undertaken in order to explain an observed phenomena) and the dogmatic inclusion of assumed starting conditions.

For all this, however, Wright was partly on the right track in his interpretation of the observed phenomena of the Milky Way and for his realization that gravitational attraction required that stars must be in motion. The universe, whether infinite or not, was dynamic and slowly changing. When (and again if) Wright is ever mentioned within modern astronomical texts it is because of a statement he makes about other "Divine Centers" and what lies beyond the body of stars that we can see. With respect to his favored spherical shell model, Wright argued that beyond the confines of the Milky Way shell was the domain of "outer darkness," where the damned were exiled. More importantly for our historical theme, however, Wright argues, "the many cloudy spots, just perceivable by us … may be external creations, bordering upon the known one (the Milky Way), too remote for even our telescope to reach." Here is the first mention by Wright of diffuse nebulae, and while it is difficult to interpret his thinking, Wright appears to be suggesting that these "cloudy spots" are other universes apparently centered on distinct and separate Devine Centers.

In modern terms Wright is almost invoking the idea of a multiverse [5]. Theologically such daring notions would not, at that time, be well received. Indeed, writing anonymously in 1755, philosopher Emmanuel Kant (working from a lengthy book review rather than Wright's original text) misunderstands Wright's intended ideas. Incapable of believing that Wright was actually invoking the notion of individual creations, each centered upon its own divine core, Kant stumbles upon the more useful notion of "island universes," not so much as distinct and separate creations but as remote stellar systems (galaxies in the modern vernacular) within one large cosmos. With Kant we also begin to see a solidification of the idea that the fixed stars are not scattered randomly throughout space but are arranged in a dynamical and systematically ordered fashion.

At the same time that Kant was speculating on the possibility of island universes, French astronomer and deacon to the Church, Abbé Nicholas Louis de Lacaille was recently returned from his travels abroad. At the age of 37 years Lacaille had ventured to the southern hemisphere, and working from a make-shift observatory established at the Cape of Good Hope, on the southern tip of Africa, he set about observing and mapping the antipodean stars. From March 1751 to March 1753 he measured the positions and

---

[5] From *Genesis* 1:14, which would have been Wright's overriding authority, we have, on the fourth day, "Let there be luminaries in the vault of the sky."

cataloged some 9800 stars, and he made careful observations of the Moon and planetary positions. Combing the latter positions with simultaneous observations made by Jerome Lalande back in France, new estimates were made for the distance to the Moon and the Sun. His southern star catalog, *Coelum Australe Stelliferum*, was published posthumously in 1763, and it was Lacaille who proffered names for most of the then new, and the now accepted, southern hemisphere constellations. During his survey of the southern stars, Lacaille inevitably came across oddities, and these he described in a catalog, *On the Nebulous Stars of the Southern Sky*, published by the Royal Academy in Paris in 1755. Lacaille's catalog contained entries on 42 nebulous objects that he divided into three kinds. Objects of the first kind were described as being nebulae without stars; the second kind constituted nebulous stars with clusters, while the third kind were described as stars accompanied by nebulosity. Somewhat ironically, from a modern perspective, objects listed by Lacaille as being of the first kind included several globular clusters, a few open clusters and one galaxy. The remarkable nova-like star Eta Carina was introduced as an object of the third kind, and so, too, is the interstellar cloud and star formation region M8, also known as the Lagoon Nebula (Fig. 1.4).

With the publication of Lacaille's catalog we are also introduced to the Coalsack Nebula (Fig. 1.5). This remarkable dark interstellar dust cloud located close to the Southern Cross constellation Crux had long been noticed by explorers of the southern seas, but with Lacaille's description we begin to perceive a sense of its mystery. Lacaille writes, "One can yet mention among the phenomena which strike the view of those who look at the Southern sky, a space of about three degrees in extension, in all directions, which appears dark black in the eastern part of the Southern Cross. The appearance is enhanced by the vivid whiteness of the milky way that enshrouds the space on all sides." [6].

Fig. 1.4 The Lagoon Nebula (M8) is located in the constellation of Sagittarius. Located some 5000 light years away, the nebula was first described by Italian astronomer Giovanni Battista Hodierma in 1654 and is just visible to the unaided eye. Image courtesy of the European Southern Observatory, ESO

Fig. 1.5 A portion of the Coalsack Nebula. Image courtesy of the European Southern Observatory, ESO

Different, and at odds with the bright glow of the stars and bright nebulae, what is the Coalsack Nebula? Easily visible to the unaided eye, it is a dark macular on the otherwise glowing abundance that delineates the Milky Way. However, does the absence of starlight in this small patch of the sky indicate a true dearth of stars, and a literal hole in the heaven's fabric? Are we really seeing the darkness, the coal-black darkness, of space, or is there something else going on? Is this one of Newton's feared-for regions in which the balance of stars has moved beyond the impotency of Buridan's donkey—a void created by the veritable crashing of stars and the rolling of heavenly spheres?

Perhaps it is a European, Old World, way of seeing things, but not every culture has viewed the Milky Way solely in terms of the nebulous light that it exudes. Indeed, the significance of the Milky Way, in many cultures around the world and through history, has been cast in terms of its voids and dark regions. For such cultures the story of the cosmos and creation of the world is defined in terms of what is framed by the stars, rather than the actual distribution of the stars themselves. For the ancient Babylonians and the ancient Greek astronomers, in fact, right through to the modern era, the night sky is delineated by the pointillist outlines of heroes, gods and mythical beasts, their helical rising and setting being used to determine the time of the seasons and the passage of the years.

For numerous Australian aborigine clans, however, the heavens are described according to the subtle sky shadows of creatures from the Dreamtime. Rather than the deep-socket of darkness and starless void seen by early European explorers, the Coalsack Nebula, to the aborigine eye, is the head of *Tchingal*, the Emu, with its aphotic body, neck, and feet stretching through the constellations of the Acrux, Centaurus, and Scorpio. In another interpretation the Coalsack is the hiding place of *Bunya*, the possum, who has climbed to the top of a sky tree after being chased by hunters. To the ancient Inca in Peru, the Coalsack Nebula formed part of *Tinamou*, the partridge. *Tinamou* lies nestled between *Hatun llamaytoq*, the Mother Llama (her body the dark clouds of the Milky Way stretching through Scorpius, her eyes the glowing orbs of α and β Centauri), the toad *Hanp'atu* (another distinct dark cloud), and the Snake, *Mach acuay,* which extends through Vela and Pupis, its head arching towards Sirius.

The Milky Way is the optical backbone of the celestial sphere, and it winds around the world as a star-studded braid, divided and crossed, according to culture and background, by constellations and dark maculation, the yin and the yang of perception, that are the ancient shadows of distant mythology.

The Star-Gauging Siblings

Although it is difficult to dismiss the beauty of the mythical heavens created by the human imagination, the Milky Way will only reveal its true secrets by detailed examination, and indeed, it is only through the deep searching of the Milky Way with a telescope that its true structure will be delineated—its saucy looks teasing out the very weave of its fabric. When exploring the heavens with a telescope, however, two basic approaches can be taken. Either set the telescope to view a random spot and see what turns up, or systematically scan the heavens, working in a methodical fashion, covering specific and well-defined swaths of the sky.

Both methods will reveal new phenomenon within an observer's eyepiece, but the latter method does away with good fortune and makes discovery inevitable. The systematic approach, in which specific regions of the sky are surveyed, is the powerful working methodology that has long been adopted by the comet hunter. Certainly chance and good fortune govern the discovery, but repeatedly looking in the right part of the sky at the right time (the pre-dawn or post-sunset regions of the eastern and western sky) will maximize the chance of catching a new wayward comet.

Eighteenth-century French astronomer Charles Messier knew this observing rule well, and he applied it with great success, discovering in his lifetime 13 new comets. What Messier also found in the sky, however, were nebulous clouds and faint star groupings that, at first glance, might be confused with a comet. These nebulae had no specific interest for Messier, but he diligently cataloged such objects when he came across them, and by 1774 his first catalog was published in the journal of the French Academy of Sciences. Several revised versions of Messier's catalog appeared in subsequent years, with the final version, *Catalogue des Nebuleuses & des amas d'Etoiles*, seeing print in 1781. The final edition contained some 103 entries [7]. From a modern perspective, among the objects contained in Messier's catalog are some of the most spectacular sights that can be found in the night sky, and his list reveals galaxies, genuine nebulae, supernova remnants, planetary nebulae, galactic star clusters, and globular clusters (recall Fig. 1.1). This modern labeling, of course, was not available to Messier or his contemporaries, but the key point in our present narrative is that nebulae, of one shape or another, were being found in relatively large numbers by astronomers, equipped, again by modern standards, with relatively small telescopes. Messier was using a modest 4-inch refracting telescope to conduct his studies, and prior to 1785 the largest telescope in the world was the 29.5-inch diameter reflecting telescope[6] constructed by the reverend John Mitchell in Yorkshire, England.

Pushing the light-gathering power of telescopes to the limits of available materials and technology, and then using this advantage to systematically gauge the sky, was the great passion of William Herschel; and it is Herschel that takes our story of discovery in new and innovative directions. In the same year that Messier's final catalog appeared, Herschel secured a permanent place in the history of science by discovering the planet Uranus—an object that, when he first swept it up in his eyepiece, he thought was a comet. The discovery of Uranus was monumental, and it brought Herschel fame and royal patronage, with the latter importantly enabling him to concentrate on building larger and better telescopes with which to explore the heavens.

With new-found fame and money at his disposal, Herschel set about moving himself and his somewhat reluctant sister Caroline to a new home, eventually named Observatory House, in Slough. It was from there, beginning in the early 1780s, that William and Caroline set about the systematic study of the heavens. Having received a copy of Messier's catalog in 1783, the Herschel siblings became interested in finding more such objects, and by 1802 the brother-sister duo had cataloged some 2500 new nebulae. Their method of discovery was highly systematic, and consisted of using a sky-drifting technique. Accordingly, the telescope would be set to some prescribed and fixed altitude, and then, as Earth's spin carried the telescope's field of view slowly across the sky, a long ribbon of the celestial sphere could be examined. When a new nebula was sighted, William, at the eyepiece, would call out a description when it crossed the center of his eyepiece view, and Caroline would then dutifully record the time and details.

Herschel became fascinated with nebulae, and he gamily strove to understand what they were and what they were telling us about the heavens. In 1782 Herschel found his first planetary nebula, and this discovery alone resulted in a long-running struggle to understand the connection between nebulosity and stars. Did new stars form in nebulous regions, or did they simple drift into them? Cataloged as NGC 7009 (Fig. 1.6) Herschel noted that he saw, "a star surrounded by a cloud of (true) nebulosity," and he speculated that the nebula might represent a star caught in the process of assembling.

---

[6] It was in 1785 that William Herschel began the construction of what was to be his great 40-foot telescope—an instrument that utilized a 47-inch diameter mirror to capture starlight. This giant telescope was paid for with funds provided by King George III, and saw first light on February 19, 1787, when Herschel observed the Orion Nebula—object number 42 in Messier's catalog (Fig. 1.1a).

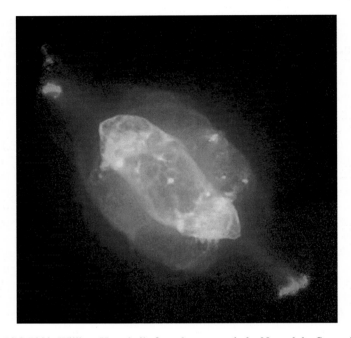

Fig. 1.6   NGC 7009, William Herschel's first planetary nebula. Named the Saturn Nebula by William Parsons (Lord Rosse) in the 1840s, NGC 7009 is located some 3000 light years away. A *cylinder-shaped* nebulosity surrounds the central star, with two jet-like streams moving away from the central region, along the system rotation axis, to terminate in two bright ansae. Image courtesy of NASA/HST

Such objects had, in fact, been observed earlier by Messier in the 1760s, but it was Herschel that coined the name planetary nebula in the mid-1780s in response to their observed round(ish) shape and smoky-green color. Although in 1784 Herschel asserted that nebulae were likely composed of some form of shining or fiery fluid, in agreement with the earlier ideas presented by Edmund Halley, he backtracked on this notion in a paper read before the Royal Society of London in 1785. Indeed, Herschel noted that wherever he looked in the zone of the Milky Way he had always found that it was composed of stars. Likewise, he observed, globular clusters could be resolved into myriad stars, and, he speculated, that they were possibly formed by gravitational clustering around a large central star. As to the diffuse nebulae, Herschel asserted that they were probably distant star clusters.

To Herschel, the stellar realm was very much a dynamic place, and he argued that the stars were initially arranged with some degree of regularity (as advocated for by Newton) but that they moved and clustered together under their mutual gravitational attraction. In addition, he noted, "there will be formed great cavities or vacancies by the retreat of stars towards the various centers," and anticipating a final universal crash of stars, he argued that the globular and open clusters might be the "laboratories of the universe, wherein the most salutary remedies of the decay of the whole are prepared." In an attempt to establish the distribution of stars in space Herschel employed a star-gauging methodology, and literally counted the number of stars that he could see in various directions around the sky.

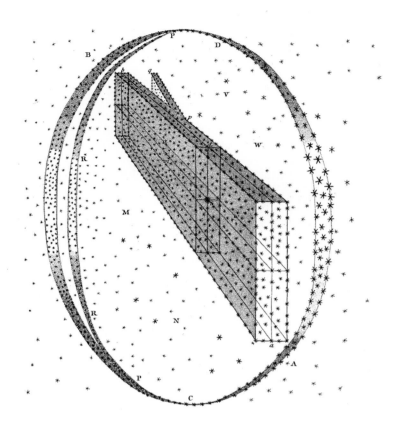

Fig. 1.7 The distribution of stars in the stellar realm as deduced by William Herschel in 1785. The stars are distributed in the form of a flattened disk, cloven in half (in the direction of Sagittarius and Scorpio), with the Sun at the center

Presenting his results to the Royal Society in 1785, Herschel argued that his gauges (to be described in more detail below) indicated that the stars were arrayed in a flattened disk, or stratum, with the Sun located at the center (Fig. 1.7). This distribution is similar to the disk-like structure evoked in 1750 by Thomas Wright of Durham, who had no more than the visual appearance of the stars on the sky to work with (recall Fig. 1.3). The stratum containing the stars, however, was not uniform but cloven—Herschel remarking that "some parts of the system indeed seem already to have sustained greater ravages of time than others." Herschel estimated that the stratum of stars had a diameter that was some 1000 times larger than the distance to Sirius, and some 100 times the distance to Sirius across. Taking, as Herschel did, the distance between the Sun and Sirius[7] to be a typical star separation distance, his estimate for the size of the stellar stratum translates to being about 9700 light years across and 970 light years in depth.

---

[7] Isaac Newton in his 1687 *Principia Mathematica* estimated that Sirius was 615,670 AU from Earth.

Fig. 1.8 The Crystal Ball Nebula. This infrared image of NGC 1514 was obtained with the WISE spacecraft, and it reveals an intriguing set of dust rings situated about the central binary star system. The nebula is located some 2200 light years away. Image courtesy of NASA/JPL—Caltech/UCLA

By the early 1790s Herschel once again changed his mind on the possible constitution of at least some nebulae. Viewing the Orion Nebula (M 42) with his newly constructed 40-foot reflector, Herschel argued that it was, "an unformed fiery mist, the chaotic material of future stars," and this viewpoint was further strengthened by his discovery, on November 13, 1790, of NGC 1514 (Fig. 1.8). Herschel described NGC 1514, now known as the Crystal Ball nebula, as "a most singular object," and he further noted that, "the nebulosity about the star is not of a star nature." In a paper read before the Royal Society of London, on February 10, 1791, Herschel discussed the topic of "Nebulous Stars, properly so called," and argued, from his extensive observational notes that some stars are truly "involved in a shining fluid, of a nature totally unknown to us." This "shining fluid," he further suggested, can even exist in space without the presence of a central star, the Orion Nebula being described as one such example where there was distinct nebulosity but no apparent connection with any specific star or stars. Herschel did note and record the four bright stars that make-up the so-called Trapezium cluster in M 42 (Fig. 1.9), but he asserted that they were unrelated to the nebulosity. It is now known that this cluster of young bright stars is responsible for illuminating most of the surrounding nebula.

Moving deeper into the 1790s Herschel began to direct his observational efforts towards objects other than nebulae—his ensuing publications relating to the study and observation of new comets, planets, planetary satellites, and stellar motion, along with the description of his new telescopes and the discovery of infrared radiation (see Chap. 2). In 1800, however, among the 30 papers he read to the Royal Society in that year, Herschel released a catalog of 500 new nebulae, to which he added some "remarks on the construction of the heavens." Indeed, from this time forward, Herschel has his sights firmly set on decoding the sky—how are the stars set out in space and what kinds of nebulous matter exist between and around the stars.

Fig. 1.9 Hubble Space Telescope view of the Trapezium Cluster at the heart of the Orion Nebula. Image courtesy of NASA/ESA

Herschel's great paper on the construction of the heavens was read to the Royal Society on June 20, 1811. It is a magnificent piece of work, covering some 67 journal pages in total when eventually published in *The Philosophical Transactions*. In this work, Herschel hammers away at the nebula—what are they, do they change, do they produce stars, or indeed, are they just dense clusters of stars that are unresolved in the telescope eyepiece? To attack the problem Herschel sets out a form of evolutionary sequence, "assorting them into as many classes as will be required to produce the most gradual affinity between the individuals contained in any one class with those contained in that which precedes and that which follows it."

Although Herschel's study plan makes sense, the scheme he develops is overly complex and not especially helpful in determining the true nature of the observed nebula. Indeed, the reader is almost overwhelmed by the excessive detail, but the end results of Herschel's efforts are clear. Through the "action of gravitation upon it" the nebulous matter in the heavens is being gradually brought together to form planets and stars.

William vacillated (to a certain extent) upon his interpretation of the nebulae. Some were composed of a "fiery fluid" in the process of condensing into stars, some were unresolved star clusters. His son John forcefully argued, during the mid- to late 1800s, that all nebulae were distant unresolved groupings of numerous stars. John Herschel followed in his father's footsteps and became one of the leading astronomers of the late nineteenth century. His opinion carried weight. Extending his father's work on star mapping and nebula-bagging to the southern hemisphere during the 1830s, John eventually published his *General Catalog of Nebulae and Clusters of Stars* in 1864. The catalog[8] contained details on some 5079 nebulae, and though he was no doubt conscious of contradicting his father's opinions, John was much more cautious in his interpretation of what the nebula clouds might actually be. The situation was further exacerbated by the 1845 claim by William Parsons (Lord Rosse) that he had resolved the Orion Nebula into individual stars. If this was the case (and indeed we now know it is not), then Orion and presumably the other diffuse and glowing nebula must be located at incredibly remote distances from us. The universe must in accordance be stupendously large, and some of the nebula must surely, in addition, be "island universes" in their own right.

It All Begins with M1

It is a question repeated in whispered reverence in rooms and meeting halls around the world: "Have you completed the marathon yet"? This is not a question of athletic prowess but rather visual acuity, commitment, and endurance. At stake is an amateur astronomer's pride and reputation. The marathon concerns the successful sighting of Messier's objects, all 110 of them [7], as they are arrayed across the ever-shifting night sky. To complete the marathon is a right of passage for the new astronomer, and just as the Boston, London and Sydney marathons begin with the first step, so, too, the Messier marathon begins with a viewing of M1.

When he discovered what we now know as M1 Messier was a 28-year old aspiring astronomer, then in the employ of Joseph Delisle, director of the Observatory of the Navy. Located in Paris, Messier was making telescopic observations of the sky from the Hotel de Cluny, and on the night of August 28, 1758, he had two objectives in mind. The first was to observe the comet newly discovered[9] by de La Nux (comet C/1758 K1), and the second was to scan the constellation of Taurus for signs that would indicate the first predicted return of Halley's Comet. Fortunately comet de La Nux was also located in the constellation of Taurus, and so having successfully swept up this new cosmic interloper Messier began to systematically sweep the constellation for additional objects. One can only assume that Messier's heart must have missed a beat when, to his surprise, his telescope eyepiece revealed a faint luminous glow: "a whitish light, elongated like the candle flame without any stars."

---

[8] A supplement to Herschel's catalog was published by John Dreyer in 1878. The *New General Catalog* (NGC) was additionally compiled by Dreyer in 1888, with further supplements appearing in the *Index Catalog* (IC) in 1895 and 1908. The NGC catalog describes 7840 objects, with the IC documenting a further 5386 nebulae.

[9] Although C/1758 appears to have been independently discovered by Messier, his first recognized new cometary *catch* occurred some 2 years later, with his detection of C/1760 B1 (Messier). Over the ensuing 38 years, 12 additional new comet discoveries were to follow.

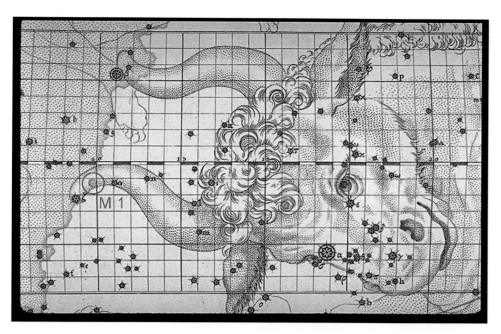

Fig. 1.10 Depiction by John Bevis of the nebula (NGC 1952, M1, Crab Nebula) in the horn of Taurus. From the *Atlas Celeste,* first published in 1786

It must have crossed Messier's mind that he might have espied Halley's fabled comet, returned early, or indeed, that he had discovered an entirely new comet, but any initial excitement was soon to be crushed. Subsequent observations revealed that the newly discovered comet-like glow moved with the stars, at a sidereal rate, rather than against the backdrop of stars, as a comet must surely do. He had found a nebula—not Halley's Comet but one of his lucid spots. Realizing that these small, faint, glowing clouds could easily confuse ardent comet hunters, Messier began to catalog their locations and appearances with the first entry in his catalog, M1, being the nebulosity found in August of 1758—the rest, as the saying goes, is history.

It is not remotely known who might have first seen Messier 1 through a telescope, but it is known that it was observed and recorded by British doctor John Bevis in 1731. The faint nebulosity appears as a rather nondescript, non-labeled small circle in the horn of Taurus (Fig. 1.10).[10] The name Crab Nebula was first applied to Messier 1 by the 3rd Earl of Rosse in 1844, the name being applied according to the view presented at the eyepiece of his 24-inch telescope located at Birr Castle in Ireland.

As with many, if not nearly all, astronomical objects named according to specific animals, the resemblance of the nebula to an adult crab is partial at best; indeed, Rosse's initial sketches indicate something looking more like a crab larva, or even a pineapple.

---

[10] Although M 1 was recorded by Bevis in 1731 and marked on copper plates engraved between 1748 and 1750, the actual sky charts were not published until 1786, at which time Bevis's posthumous *Atlas Celeste* appeared.

Fig. 1.11 The Crab Nebula, in all its glory, as revealed by the Hubble Space Telescope (also see Fig. 2.19 later in this book). Image courtesy of NASA/HST

Viewing the nebula 4 years later with his giant 72-inch telescope, Rosse even admitted that he could no longer see any resemblance in the nebulous cloud to a crustacean. But the name, for all of its inexactitude, has stuck (Fig. 1.11).

Although naming an object begins the societal process of assimilation, the exact nature of the Crab Nebula was not to be fully realized until the 1940s had run their course. Indeed, the next important step in the story of M1, after its naming, was the realization that it was an object with a well defined birth date. From our terrestrial viewpoint, the Crab Nebula came into existence on July 4, 1054.

Evidence supporting the relatively recent cosmic age of M1 was provided on two fronts—one involving a combination of history and theoretical reasoning, and the other based on direct observations. First, writing a series of research papers that appeared (mostly) in the *Publications of the Astronomical Society of the Pacific,* starting in 1921, the great Swedish astronomer Knut Lundmark, then at the Lick Observatory in California, set out to investigate the distribution of historically observed nova, literally "new stars," in the sky. Likewise in 1921 John Duncan, at the Mount Wilson Observatory, was able to measure changes in the physical size of the Crab Nebula—it was getting bigger. Let us deal with Lundmark's papers first.

Throughout history novae have been both a marvel and a mystery. What exactly was going on so that a new star might form and then eventually fade away? Such behavior was at odds with the supposed constancy of the heavens. By the early twentieth century, however, a number of mechanisms had been proposed to explain the appearance and generation of novae. Perhaps they were caused by a star chancing to pass through a cloud of gas, a frictional effect that then caused the gas to briefly glow. In this case, it was envisioned that novae were literally massive shooting stars. Perhaps they were new stars—the material gathering to produce the star literally glowing during the final accretion formation phase. Perhaps novae outbursts were produced when a wayward planet crashed into the surface of its parent star, or perhaps they were produced by the collision of two stars in space.

Writing to the *Publications of the Astronomical Society of the Pacific* in July of 1921 Lundmark set out to describe the apparent distribution of novae in the Milky Way. His first task was to shoot down the idea that the probability of seeing a nova in any given region of the sky was related to the star density in that area. Indeed, Lundmark argued that novae were more likely to be found in places such as "the borders of the galaxy [the band of the Milky Way] or of the so-called coal-sacks." With these observations in place Lundmark suggests that perhaps novae were "related with the nebular matter in our universe," and he also noted that the novae tend to be frequently observed in the Sagittarius-Scorpius direction—the direction we now know to be associated with the galactic center.

In October of 1921 Lundmark again wrote to the *Publications of the Astronomical Society of the Pacific*. In this second paper he considered the historical records relating to the observation of novae, and it was in this publication we find the very first suggestion that the Crab Nebula might be related to a new star observed in 1054. Certainly, however, Lundmark understated the relationship, and he simply added a footnote to one of his tables that the nova of 1054 was, "near N.G.C. 1952."

In a third lengthy paper submitted to the *Publications of the Astronomical Society of the Pacific* in early 1923, Lundmark once again sets out to review "some facts" relating to novae. He again analyzes the sky distribution of novae, finding, as with his 1921 papers, a distinct concentration in the direction of Sagittarius and Scorpio; he added, however, some support to the idea that the distribution of novae was similar to that of the Wolf-Rayet stars,[11] although he continued to press forward that the idea that novae are produced by stars entering into "a nebular territory of the universe." By the mid-1920s Lundmark had begun to move on to topics further afield, pioneering the study of the "anagalactic nebulae," or as we would now call them, distant galaxies. Indeed, the 1920s were a remarkable time with respect to expanding our understanding of the universe and the distribution of stars.

With John Duncan, at the Mount Wilson Observatory, we return to the story of M1. Writing in the *Proceedings of the National Academy of Sciences* for May 1921, Duncan discussed changes observed in the Crab Nebula in Taurus. Building upon the results published by Carl Lampland working at the Lowell Observatory, Duncan compared two images of the Crab Nebula taken with the 60-inch Mount Wilson reflector in October of 1909 and April 1921. Noting that, "it was seen at once that changes … had occurred," Duncan found that the material in the nebula had expanded by some 1.54 arc seconds in 11.5 years. These numbers combine to produce an angular expansion rate of 0.13 arc

---

[11] Odd among stars in general, the Wolf-Rayet stars show strong emission lines in their spectrum (see Chap. 2). First described by Charles Wolf and Georges Rayet in 1867, the Wolf-Rayet stars are now known to be pre-supernova stars.

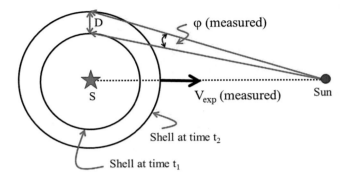

Fig. 1.12 Simplified geometrical method for determining the distance to an expanding nebula. Direct observations provide values for the angular expansion φ, in time interval T, and the
expansion velocity *Vexp* [8]

seconds per year. The expansion of the nebula gas was evident to Duncan, but he was
not able to deduce anything about the actual expansion speed. Assuming that the expansion speed were 25 km/s (a speed in fact some 60 times smaller than the actual value),
Duncan concluded that the distance to the nebula was 100 light years (again, a factor
that is about 65 times too small in comparison to modern-day values). At this stage the
situation and status of M1 remained in stasis for a number of years, and the story only
picks up again in the late 1930s.

The first photographic spectrum of the Crab Nebula was recorded by Vesto Slipher in
1913, but it was not until a study by Nicholas Mayall, in 1937, that a Doppler shift velocity
(see Chap. 2 for details) was to be obtained. Mayall found a staggering expansion velocity
of 1300 km/s. Importantly, with the expansion velocity measured, Mayall could determine
for the first time an accurate distance for the Crab Nebula. The method applied to determine the distance to an expanding nebula is illustrated in Fig. 1.12, and it was through
such an analysis that Mayall deduced a distance of 1500 pc (some 4900 light years). With
the distance, expansion velocity, and angular diameter of the nebula known it is then a
straightforward step to determine how long the nebula has been expanding—the answer
came out to about 800 years. At this point in his analysis Mayall drew attention to
Lundmark's previously suggested linkage of the Crab Nebula to the new star observed in
1054 and comments that the association was now "strongly" supported.

Mayall's paper of 1937 essentially sealed the deal, but the defining paper, in which the
Crab Nebula is firmly linked to the supernova of 1054, is that by Mayall and Jan Oort.
Appearing in the *Publications of the Astronomical Society of the Pacific* for February
1942, the paper by Mayall and Oort sets out to refine the observational data and establish
an absolute magnitude for the historically observed nova of 1054. Concluding that, given
an observed brightness well in excess of that of the planet Venus in the night sky, the absolute magnitude must have been of order $M = -16.5$ (see Appendix at the end of this book).
Mayall and Oort confidently remarked at the end of their paper that, "The Crab Nebula
may be identified with the 1054 supernova."

The word supernova is used by Mayall and Oort in the title of their research paper, and this reflects a new development in the study of star systems. The term "super-nova" was introduced by Walter Baade and Fritz Zwicky in an article published in the May 1934 issue of the *Proceedings of the National Academy of Science*. In this six-page article Baade and Zwicky distinguished between the "common novae" and the brighter, less numerous "super-novae," for which in the latter case they argue, "the phenomenon…represents the transition of an ordinary star into a body of considerably smaller mass."

The only supernova candidate that Baade and Zwicky offered for our Milky Way Galaxy was that observed and described by Tycho Brahe in 1572. Zwicky, along with Rudolph Minkowski, went on to develop a classification scheme for supernovae in the early 1940s, and subsequently it has been realized that novae[12] and supernovae have a range of distinguishing spectral characteristics, production mechanisms and progenitor star systems. The important role of supernovae in shaping the properties of the interstellar medium, giant molecular clouds and inducing star formation will be explored in later chapters. For the present, however, the key idea behind the supernovae phenomenon (of whichever form) is that the material that was once incorporated into a star, as Baade and Zwicky argued, is rapidly expelled into space. In the process vast quantities of radiative and kinetic energy are released into the surrounding interstellar medium, stirring and twisting it into a chaotic flows and shock-heated domains.

Importantly for our very existence, supernovae are the great engines of material and atomic transformation. It is through supernovae explosions that the vast reservoirs of hydrogen and helium, the original matter produced within the first few minutes following the primordial Big Bang, are being slowly converted into other chemical elements. Astronomer Carl Sagan is famed for noting in his wonderful book *Cosmos* (first published in 1980) that, "we are made of star stuff," and he was entirely right. All of the vital chemical elements that allow life to exist, planets to form and evolution to work were formed in supernovae.

English-born astronomer Geoffrey Burbidge is rightly famed for his part in the production of the renowned and trail-blazing B^{2}FH paper (see Chap. 5), published in the journal *Reviews of Modern Physics* in 1957, and it is said that he once remarked that modern astronomical research is equally divided between the study of the Crab Nebula and the study of everything else. He had a point, even if overstated, and the Crab is now a prominent visual icon of astronomy. Not only was the Crab Nebula the first nebula to be recognized as a supernova remnant; it was the first radio (in 1949) and X-ray (in 1964) emission source to be recognized in the sky (after the Sun, that is). Additionally it is the only historically observed and known supernova remnant to be associated with a pulsar—as recognized by observers using the Aricebo radio telescope on November 9, 1968. The first pulsar (PSRB1919+21) had only been identified a year earlier by astronomers at Cambridge University in England, in November of 1967. The Crab pulsar is additionally the only pulsar to ever have its brightness variations recorded at optical wavelengths. Indeed, as an object of curiosity, investigation and marvel we shall meet the Crab Nebula again and again in our ensuing narrative (see especially Chap. 2 of this book).

---

[12] The phrase 'common nova' is now no longer in use, and the hyphen in the original spelling by Baade and Zwicky has also been dropped.

The Columns Arrayed Before Us

When I heard the learn'd astronomer;
When the proofs, the figures, were ranged in columns before me;
When I was shown the charts and the diagrams, to add, divide, and measure them;
When I, sitting, heard the astronomer, where he lectured with much applause in the
*   lecture-room,*
How soon, unaccountable, I became tired and sick;
Till rising and gliding out, I wandr'd off by myself,
In the mystical moist night-air, and from time to time,
Look'd up in perfect silence at the stars.

—Walt Whitman

As a "learn'd astronomer" these past many decades this author begs to differ, at least on this one issue, with Walt Whitman. The way to *know* the stars is through measurement, and by the very nature of measurement the best way to present the information it contains is through figures, charts, diagrams, columns and proofs. True, such arrangements lack the mystical beauty that Whitman was aiming for, but they offer the yet greater beauty of a true understanding. To *understand* the stars, what they are and how they are distributed in space, measurements of quantifiable parameters are essential. There is no other way in which the true properties of the stars can be deduced. By making such blunt statements about the stars, we do not wish to diminish human imagination, and we certainly hope that you will not become "tired and sick." The point, however, is that if we are to say anything that is actually meaningful about the incredible universe in which we reside then a scientific approach is by far our best approach, and this requires that we have numbers associated with physical quantities to work with.

Making observations and measuring physical properties are certainly part of the process of doing science, but without a theoretical model to place them within some well-defined context, they essentially tell us nothing. Indeed, it is the great power behind the scientific method that the theory is used to interpret and understand the observations at the same time that the observations are used to refine the theory. Science is an iterative process, continually refining the theoretical predictions against new and deeper sets of observations.

By way of an example of how this process works, let us look at the predictions associated with William Herschel's star gauging (recall Fig. 1.7)—that is, what does the act of physically counting the stars tell us about the actual distribution of the stars? To begin with, as Herschel did before us, we have to make a number of assumptions, and then, based on our assumed theoretical model for the distribution of stars, we shall make a prediction about what the actual star count data will show. In this manner the consequences of our assumptions can be tested. If, and only if, the model prediction agrees (within reasonable limits) with the observations can the model be accepted, indicating that we have a reasonably good handle on what the observations are actually telling us. What Herschel assumed, and reasonably so for the time, were the following conditions:

- The stars were uniformly distributed in space in essential accord with the uniform spacing idea advocated by Isaac Newton.

- His telescope was able to 'reach' to the very limit of the stellar system (that is, he was able to observe to the 'edge' of the universe).
- All stars have the same intrinsic luminosity.

The first assumption dictates that the number of stars per unit volume of space is constant. In this manner, the number of stars $N(R)$ within a sphere of radius R about the Sun will vary according to the radius cubed: $N(R) \sim R^3$. The third assumption is important since, for a given luminosity, the brightness of an object decreases according to the inverse square of its distance from the observer: $B(R) \sim 1/R^2$. From this latter relationship we obtain the useful result that the farther a star is away from us, the fainter it will appear, or, alternatively, the fainter a star is observed to be, the further away it must be from us. Herschel introduced the third assumption for two reasons. First, it allowed him to make some progress in interpreting the observations, and second he didn't know the true intrinsic variation in stellar luminosity. By assuming that all stars have the same luminosity Herschel was then able to argue that the fainter the stars he could see in a given direction, the greater the depth of his line-of-sight into the stellar system must be. That is, the star system extended furthest in those regions where he counted the greatest number of faint stars.

Having introduced these three assumptions about the characteristics of stars and how they can be observed, the next step is to make a prediction that is testable and can be compared against the observations. One way of doing this is to consider what happens if we make two star count surveys, one out to a distance R_1 from the Sun and the other to a distance R_2 greater than R_1 from the Sun. In the first survey (to radius R_1), N_1 stars are counted, and in the second survey (to distance R_2), N_2 stars are counted. Accordingly, it turns out, if the faintest star visible in the deeper survey to distance R_2 is 1 magnitude fainter than the faintest star in the survey out to distance R_1, then the total number of stars counted in the deeper and fainter survey should be four times greater than the number counted in the closer survey. This result can be written as $(N_{m+1}/N_m) \approx 4$, where m corresponds to some set magnitude [9]. Importantly, this is actually a result we can test against the star-gauging observations.

An Evening with Norton

Counting is easy; it's as simple as $1+1+1=3$, and numbers, as Galileo told us, now long ago, are the "words" of nature. Numbers are also the stock-in-trade of the scientist, and within the act of counting resides the discovery of the universe. Pliny the Elder, in his *Natural History* records that Hipparchus (circa 135 B. C.) was inspired, so it is claimed, to count the stars and construct his famous star catalog on the basis of seeing what we would now call a supernova—a new star that suddenly appeared, bright and bold, where no star had been seen before. Such new stars do not appear very often.[13] Just three have been witnessed from within our galaxy since 1054. But when they do appear they can dominate the sky.

---

[13] Many supernovae have been seen in other galaxies, and the inferred supernova rate for a galaxy like our own is of the order one every 50 years. The fact that the observed rate is on average about 1 every 350 years is entirely due to the effects of interstellar dust—as will be described later in Chap. 3.

To the ancients they were odd, unexpected and a complete mystery, inexplicably offering something altogether novel in the sky.

When a (galactic) supernova is seen the total number of visible stars has increased by just one, though a temporary one, but the questions, thoughts and ideas that follow such events can, from a human perspective, be profound. On seeing and then thoroughly investigating the supernova observed in 1572 the great Tycho Brahe suggested that it was literally a new star formed from nebulous material such as that seen in the Milky Way. Following his own intensive investigation of the supernova observed in 1604, Johannes Kepler suggested, like Brahe had before him, that it had formed from the same kind of nebulous material that was seen in the Milky Way.

Such ideas are now known to be wrong (supernovae actually represent an explosive end phase of stellar evolution rather than a formative one), and upon reading Galileo's *Siderius Nuncius,* Kepler seemingly recanted his ideas about stars forming from nebulous material. Nonetheless it is the case that, from the early seventeenth century onward, the radical idea that new stars might actually form and grow within the stellar realm took hold. The story of star formation will be taken up later in Chap. 5, but for the moment we return to the "fixed" stars.

In order to test the prediction derived above, that the number of stars counted will increase by a factor of 4 every time we increase the survey limit by one magnitude, we can either, as Herschel did, turn to the telescope, or we can employ the indoor astronomer's standard dodge and spread out a printed star map. Indeed, the procedure to be employed in testing the prediction on star counts is the same whether we work at the telescope or sit at a desk: count the number of stars in each of six magnitude categories, in several different regions of the sky, and see what nature actually gives us. Here a star atlas study will be used, and the results obtained can, in fact, be reproduced by anyone who has a few hours to spare.

The hard and time-consuming work, of course, has already been done by the observer and cartographer in the production of the atlas. Of its kind, *Norton's Star Atlas*, as first published by Arthur Norton in 1910 and now into its 20th edition, is one of the best around. Its maps are clear and crisply presented, and it is complete down to as faint as magnitude +6 stars. These latter stars are about as faint as the unaided eye can see on a good clear night.

Rather than count every star on every map, which would not be fun, the author has counted only those stars in a random selection of 24 regions, with a total area corresponding to a third that of the entire sky, were considered. This is a purely practical step reflecting the time constraints of a person's life. The task of counting took just a few hours to complete (rather than the months and years spent by Herschel at the telescope), and the results are presented in Table 1.1.

Table 1.1 The author's results from the star counting exercise based on 24 selected regions in *Norton's Star Atlas*

Magnitude m:	1	2	3	4	5	6
$N(m)$:	7	18	59	174	650	1695
$N(m+1/m)$	–	2.57	3.28	2.95	3.73	2.61

The area surveyed in this star count corresponds to about a third of the entire sky

It is immediately obvious from Row 2 of Table 1.1 that there are many more faint stars than bright ones, as, indeed, we would expect just from a casual glance at the night sky. Given that the total area of the search maps analyzed corresponds to 33.5% of the entire sky, the total number of stars in each of the magnitude categories is expected, all things being equal, to be 3 times the results listed in the Row 2 of Table 1.1. The so-called *Bright Star Catalogue*, edited for many years by the indefatigable Dorrit Hoffleit of Yale University and first compiled in 1908, indicates that there are a total of 22 stars brighter than magnitude 1 in the sky and 8692 brighter than magnitude 6. From Row 2 of Table 1.1 we estimate that there should be 21 stars brighter than magnitude 1 and 7809 stars brighter than magnitude 6. Our overall numbers are not perfect but in reasonably good agreement with the *Bright Star Catalogue* (within 5% at magnitude 1 and 10% at magnitude 6), and this gives us some confidence in our methodology and counting ability.

What is immediately clear from Row 3 of Table 1.1 is that our prediction concerning the star count numbers relating to surveys having limiting magnitudes that differ by one, i.e., $(N_{m+1}/N_m) \approx 4$, is wrong. Indeed, the observational results indicate $(N_{m+1}/N_m) \approx 3.0$. Hugo von Seeliger, at Munich University, in a highly mathematical paper published in the journal *Astronomische Nachrichten* in 1909, conducted a star count analysis (similar in scope to the Norton one just described) on the *Bonner Durchmusterung* data,[14] and found that the number of stars increased in a ratio that varied from 2.8 to 3.4 per magnitude jump—a result comfortably close to our much more restricted analysis. In short, however, the observed star count data is simply not consistent with the prediction, and this immediately tells us that the assumptions upon which the prediction is based must be wrong. Well, such are the strictures of the scientific approach, and what we need to do now is throw away our original set of assumptions and rework our initial ideas.

Jumping ahead to the modern era, it is easy to identify the problems with our initial set of assumptions, and in fact, none of them is actually true all of the time. Stars are not in general uniformly spread throughout space, they are not all of the same luminosity and no telescope can see (detect) every star to the very limits of the stellar realm. This being said, however, the number of stars per unit volume of space, in the region out to at least a few hundred parsecs from the Sun, is approximately constant. This was determined by the Research Consortium on Nearby Stars (RECONS) group of astronomers.

The key point about the stars included in Fig. 1.13 is that they have well determined parallax distances (see below) and well determined luminosities. The data counts, however, are not complete, and as one attempts to complete surveys at larger and larger distances from the Sun, so the more stars will be missed. The number counts are believed to be complete, that is, every star is accounted for, out to a distance of 5 pc from the Sun—a volume of space that contains 47 stars. At greater distances than 5 pc from the Sun the so-called under-sampling problem begins to become more and more important.

---

[14] The *Bonner Durchmustering* all-sky star catalog was compiled at Bonn Observatory between 1859 and 1903. The final catalog provides data on the position and apparent magnitude of 325,000 stars down to a limiting apparent magnitude of +10.

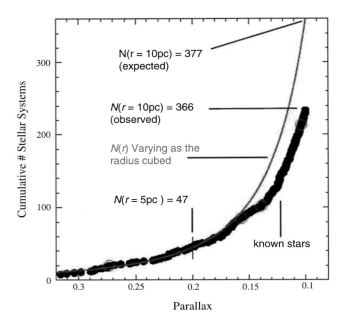

Fig. 1.13 The total number of known stars versus distance out to 10 pc from the Sun. The star count is complete out to a distance of 5 pc from the Sun. Beyond this distance the under sampling issue becomes increasingly important. Diagram from data supplied at RECONS.org

Using the star data contained within the 5-parsec region surrounding the Sun, it was found that the number of stars within a given region of space does indeed increase, to a reasonably good approximation, in lock-tight step with the volume, that is, $N(r)$ does increase as the radius cubed, and specifically can be written mathematically as $N(r) = 0.377$ $r(\text{pc})^3$. This result is equivalent to saying that there are 0.09 stars per parsec cubed volume of space. The RECONS group indicates that out to a distance of 10 pc from the Sun there are a total of 366 known stars (including the Sun and white dwarfs), while the expected number, from the complete 5 pc sample, is $N(r = 10 \text{ pc}) = 377$. The under-sampling, that is, the number of stars not yet accounted for out to a distance of 10 pc, assuming stars are uniformly spread out in space is, therefore, 11.

Up to a point, then, Newton's idea of uniformly distributed stars does hold true, but it is not true for all spectral types of stars and/or over very large volumes of galactic space. We will come back to discuss this issue later on.

That stars have very different intrinsic luminosities was certainly something that William Herschel suspected might well be true. Indeed, Herschel knew from his observations of binary star systems that stars, although at the same distance from us, can have very different brightnesses, and that this observation can only be accounted for if the component stars have different luminosities. This result, in fact, became immediately evident as soon as the first really believable parallax distance measures for three nearby stars were published circa 1838.

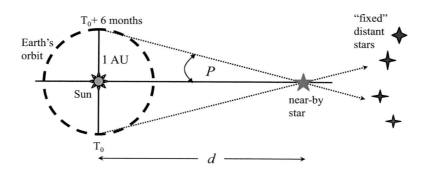

Fig. 1.14 The essential geometrical basis of stellar parallax. The key idea is to use the diameter of Earth's orbit around the Sun to induce a change in the apparent position of a nearby star relative to much more distant 'fixed' stars

Parallax is of paramount importance in astronomy, and it is the first rung on the cosmic distance-scale ladder. The geometrical idea behind parallax (Fig. 1.14) is simplicity itself, but the practical measurement of the angle of parallax is fraught with difficulties and complications. The essential baseline for a parallax measurement is that of Earth's orbital diameter, with positional measurements of a given star being made 6 months apart. Any shift in the apparent position of a nearby star, as measured against the backdrop of much more distant stars and/or galaxies, can accordingly be determined and converted into a distance.

By definition, if the angle of parallax, which is actually half the star's measured angular shift over the 6-month time interval, is 1 second of arc, then the distance is 1 pc. Indeed, if the angle of parallax P is expressed in units of arc seconds, then the distance d to a star, in parsec units, is simply $d(pc) = 1/P(arcsec)$.

Astronomers have been trying to make reliable parallax measurements for most of the history of astronomy. It was the Holy Grail and El Dorado all rolled into one. With a parallax measurement you have a physical distance, and this combined with flux measurements tells you about the physics, motion, and spatial distribution of the stars. The ancient Greek philosophers knew all about the geometry of parallax but lacked the technology to measure small angles accurately. Tycho Brahe, in the late sixteenth century, knew all about the importance of parallax, and though his technological reach far exceeded that available to the ancients, he nonetheless failed to find any bi-annual shift in the position of the stars. Indeed, with this result in hand, Brahe rejected the Copernican hypothesis, arguing that if Earth was truly in motion around the Sun, then the stars would betray this fact.

Brahe was correct in his interpretation of the available data, but what he failed to appreciate was that the stars are very, very much more distant from the Sun than the region located just beyond the orbit of planet Saturn. Brahe's view of the universe made it an extremely compact, even claustrophobic, one by modern standards.

Robert Hooke, in the early eighteenth century, invented the zenith telescope, and built it into the framework of his home, from ground floor through to rooftop, to specifically look for the parallax of one specific star, γ Draconis. He failed in his search, but he opened the doorway to success by others. First, James Bradley used another zenith telescope to discover stellar aberration and terrestrial nutation, in the late eighteeenth century, and then in

1838 Friedrich Bessel successfully determined the parallax for the binary star 61 Cygni, "the flying star," so-named for its large proper motion. This system has (in modern numbers) a parallax of 0.28588 seconds of arc, and this places it at a distance of 3.498 pc from the Sun.

In short order after Bessel's publication, Friedrich Struve, at the Pulkovo Observatory in Russia, announced his findings for the parallax to Vega, the brightest star in the constellation of Lyra. And likewise, Thomas Henderson announced his measurement for the parallax of α Centauri. These latter two stars are located at distances of 7.679 and 1.325 pc, respectively.

Here, within these numbers, is the answer to the question concerning stellar similarity. The closest star on the initial parallax list (after the Sun), α Centauri, is not the brightest star in the night sky. Indeed, of the stars observed, Vega, some 5.8 times further away from the Sun than α Centauri, was the brightest. This can only mean one thing—Vega is intrinsically more luminous than α Centauri. In contrast the faintness of 61 Cygni indicated that it must be intrinsically less luminous than either α Centauri and Vega. Clearly, therefore, stars are not identical objects, since they have different intrinsic luminosities.

The closest actual star to the Sun, at the present epoch, is Proxima Centauri, which was discovered by Robert Innes in 1915. Dutch astronomer Joan Voûte provided a first parallax distance to Proxima in 1917, and the presently accepted value gives a distance of 1.302 pc— making it about 4744 AU closer to us than α Centauri. In terms of luminosity Proxima Centauri radiates into space just 1/900 that of the energy radiated into space by α Centauri.

From a human perspective, perhaps the most remarkable thing about our nearest stellar neighbor is that we cannot see it with the naked eye. Indeed, a relatively large telescope (and a detailed star chart) is required to bring Proxima Centauri into view. This remarkable circumstance is further multiplied by the fact that Proxima additionally represents one of the most common kinds of stars to populate the galaxy. In terms of an observational selection effect, the unaided human observer can see not one example of the most common type of star that there is.

With the distances to three stars accurately known (four if one includes the Sun) a study could be made, for the first time, of comparative stellar energy output. By 1838, spectroscopy had already revealed a temperature difference between the stars, and in modern terms the spectral types (which is a proxy measure of temperature—see the Appendix in this book) of 61 Cygni A, Vega, α Centauri A, and the Sun are K5, A0, G2, and G2, respectively. This makes Vega the hottest (at 9600 K), and 61 Cygni A the coolest (at 4500 K). To determine how much energy these stars radiate into space, which is a direct measure of how much energy they generate by fusion reactions within their interiors, a measure of both the energy flux received at Earth and distance is required. With the distances now determined by parallax measures, however, the luminosity of 61 Cygni A, the Sun, α Centauri A and Vega can be compared and, in solar units, the values are 0.15, 1.0, 1.52 and 40.12, respectively. The energy output of our first four comparison stars varies by a factor of 267, and continued research has revealed that stellar energy output (on the main sequence—to be described more fully in Chap. 5) varies by some 10 orders of magnitude, with luminosities varying from 10^{-4} to 10^6 times that of the Sun.

Well, so much for the constant luminosity condition assumed by William Herschel in his star gauging. From our present position it is now abundantly clear that stars have a whole range of luminosity values, and that distances cannot be gauged purely on the basis of perceived brightness. Herschel was aware, to some extent, of this limitation in his analysis, but he would probably have been greatly surprised by just how much stellar energy output can vary from one star to the next.

Herschel's final assumption that his telescope could see to the edge of the stellar distribution might at first seem entirely reasonable. This assumption, it turns out, however, is most definitely wrong, and the reason for this is interstellar dust. The dust component of the interstellar medium is both clumpy and all pervasive. At some level the dust is everywhere within the disk of the galaxy, even when the view looks clear. It is the interstellar dust, and the way in which it absorbs and scatters starlight, that limited the view available to Herschel. Contrary to his belief, Herschel could not see to the edge of the stellar realm. Indeed, in the galactic plane Herschel could only see as far as the interstellar dust would allow, and this, it turned out, was not very far.

The interstellar dust, as we shall see in more detail in Chap. 3, acts to make distant stars fainter than they would otherwise be. This dust-dimming effect, combined with his assumption of constant stellar luminosity, severely affected Herschel's scale-reckoning for the stellar realm (recall Fig. 1.7), and it also pushed the brightness of the stars well below the limiting magnitude of his telescope after a distance of just a few thousand light years from the Sun. Herschel knew nothing of interstellar dust and its starlight dimming effects, and nature effectively tricked him into deducing a size to the stellar realm that is almost claustrophobic by modern-day standards.

Further Revelations from Norton

Herschel removed the speckled tent-roof from the world and exposed the immeasurable deeps of space, dim-flecked with fleets of colossal suns sailing their billion-leagued remoteness.

—Mark Twain
(The Secret History of Eddypus, the World-Empire)

To this point we have only considered the consequences (and pitfalls) of star counting. At this stage we begin to broaden our investigation and take into account both number and direction. Here the idea is to begin analyzing the observed locations and the clustering of stars and nebulae in order to learn something about their distribution in space as well as their distribution with respect to each other. Yet again, *Norton's Star Atlas* will provide us with a dataset.

In order to look at the spatial distribution of objects in the sky, we first need to construct an appropriate map. To do this the celestial sphere will need some unraveling, taking its 3-dimensional form to a 2-dimensional projection that can be printed on a page. Figure 1.15 illustrates the essential characteristics of the celestial coordinate system in which the location of an object is described according to its right ascension and declination. The right ascension (RA) is measured in units of hours along the celestial equator, with 1 hour corresponding to an angular displacement of $15°$. The declination (or δ) is the angular measure directly north or south of the celestial equator. For the maps shown here imagine peeling back the surface of the celestial sphere to produce a rectangular star map strip (Fig. 1.16), and since it is not critical to our specific analysis do not worry about any wrinkles or distortions that the peeling off process must technically produce.

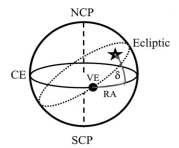

Fig. 1.15 The right ascension (RA) and declination (δ) coordinate system. The zero point of the RA measure is set according to the location of the vernal equinox (VE). The ecliptic defines the plane of Earth's orbit around the Sun, CE indicates the *circle* of the celestial equator, while the *dotted line* indicates the direction of Earth's spin axis, which intercepts the celestial sphere at the north and south celestial poles (NCP & SCP), respectively

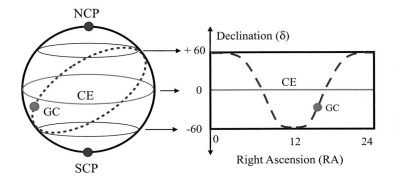

Fig. 1.16 Peeling-back the celestial sphere to produce a rectangular map of the heavens. Here it is just the region between −60 and +60° declination that is being unraveled. The *dashed circle* corresponds to the mid plane of the Milky Way, and GC indicates the direction to the galactic center. The *curved line* shown on the 2-dimensional strip corresponds to the unwound development of the Milky Way

Earlier, the star count data was extracted from 24 randomly selected regions on the sky, and it is these same 24 regions that we shall use to investigate the spatial distribution of stars, globular clusters, galactic clusters, spiral nebulae (galaxies), and diffuse nebulae (recall Figs. 1.1a–d). Figure 1.17 indicates the right ascension and declination locations of the 24 selected regions from within *Norton's Star Atlas,* over which the count and spatial data will be compared. The location of various selected Messier objects and notable star and galaxy clusters is also revealed in Fig. 1.17. Figure 1.18 shows the directional star count data, where the total number of stars brighter than apparent magnitude +6 is indicated in each of the 24 selected regions.

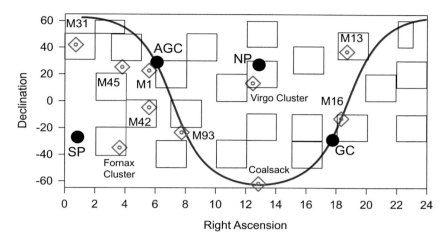

Fig. 1.17 Location map for the 24 selected regions from which count comparisons will be made. The location of several selected Messier objects, nebulae, and galaxy clusters are additionally shown on the map: GC corresponds to the location of the galactic center, AGC is the anti-galactic center, while NP and SP are, respectively, the north and south galactic poles

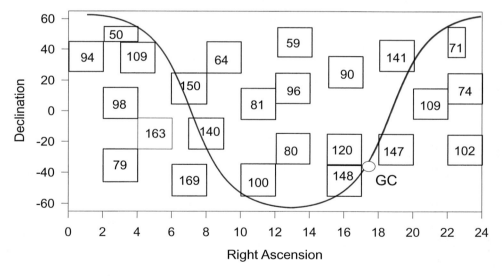

Fig. 1.18 Star counts in the 24 selected sky regions

These direction data counts, though not overly convincing, do reveal that there are more stars in those regions close to the curved line of the galactic plane, and that more stars are most definitely seen in the direction of the galactic center. As one moves towards the regions close to the galactic poles (refer to Fig. 1.17), however, the total star count numbers begin to fall. The star count data shown in Fig. 1.18 is entirely consistent with the results deduced by William Herschel in 1785 (recall Fig. 1.7). The greater numbers of stars are located in the plane and circle of the Milky Way, with fewer stars being found per unit sky area as one moves above and below the galactic plane. Typically the star count in those regions located in or close to the Milky Way is of order 150, while the counts in those regions closer to the galactic poles is about half this value and of order 80.

About the same number of stars are seen in those regions bordering the galactic center (GC), in the direction of the constellation Sagittarius, and those regions close to the galactic anti-center (near RA $\approx$ 6 hour, $\delta \approx$ +30°). This observation suggests that the Sun must be located at, or close to, the center of the stellar realm—the same result as that deduced by Herschel in 1811. The star count data from our selected 24 regions in *Norton's Star Atlas* bears out the conclusion of William Herschel. In terms of pure number counts the stellar realm does indeed appear to have the basic shape of a circular disk with the Sun situated at the center. Herschel was right, although as we shall see, the data does not actually reveal all there is to be seen. The dimensions of the disk, however, are not revealed by the star count data, and, of course, the star counts are modulated and ultimately bounded by the presence of interstellar dust (but more on this later).

Figure 1.19a indicates the galaxy and globular cluster counts in each of our 24 selected regions. We now begin to see some entirely new features. The upper number in each region of map (a) reveals the galaxy count, and clearly the greatest number of galaxies are seen in those regions close to the galactic polar regions (RA $\approx$ 12, $\delta \approx$ +30°, and RA $\approx$ 1, $\delta \approx$ −30°). No galaxies are seen in those regions that span the curve of the Milky Way—the exact reverse of the star count data (Fig. 1.18). In contrast to the galaxy counts, the globular clusters, the lower number in each region of map 19 (a) are preferentially found in those regions located on or near to the curve of the Milky Way. Not only this, however; a greater number of globular clusters are found in those regions close to the galactic center. No globular clusters are seen in those regions that are located in a direction 180° away from the galactic center.

Clearly, the distribution of globular clusters, as seen in the sky from Earth, is very different from that displayed by the galaxies. Most prominent among the galaxy count regions are those close to the north (NP) and south (SP) galactic poles. Those regions surveyed close to the NP are picking-up members from the extensive Coma and Virgo cluster of galaxies, while those regions surveyed close to the south galactic pole are sampling regions from the Fornax cluster of galaxies.

Figure 1.19b shows the count data for galactic clusters (upper number in each box) and diffuse nebulae (lower number). Again, we see different distributions from those that have gone before. The galactic clusters are only found in those regions that are close to the curve of the Milky Way, and there are approximately equal numbers of galactic clusters in all those regions intersecting the Milky Way. The diffuse nebulae are additionally only found in those regions that are on and close to the curve of the Milky Way, but now they are mostly found in the direction towards the galactic center. Like the stars, therefore, galactic clusters are a feature of the galactic disk, and it is evident that they have a different spatial distribution to that of the globular clusters. The spatial distribution of the diffuse nebula is very similar, in fact, to that displayed by the dark nebula, and it is the absorption of starlight by the latter clouds that was responsible for Herschel finding a cloven or split disk-like distribution (recall Fig. 1.7) of stars.

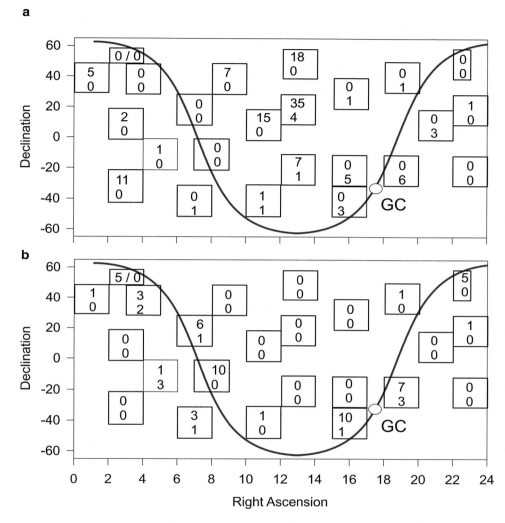

Fig. 1.19 (**a**) Galaxy (upper number in each box) and globular cluster (lower number in each box) counts in the selected regions. (**b**) Galactic cluster (upper number) and diffuse nebulae (lower number) counts in the selected regions

The asymmetry that is evident in the spatial distribution of the globular clusters and the galaxy counts is, it turns out, highly important for interpreting the location of the Sun with respect to the galactic center and in determining the distribution of interstellar dust within the plane of the Milky Way Galaxy. The hidden secrets and meaning behind the observed asymmetry of the globular clusters and galaxies will be revealed, in detail, in Chap. 3.

The Gathering of Stars

The constellation of Taurus is bold and brash, and it brings to mind images of powerful beasts and thoughts of strength, stamina, willfulness, and belligerence. There is, after all, nothing subtle about a charging bull, or a bull in a china shop. And yet, nestled within the bright framework of the stars that constitute the mythical bovine falls a softer light and a set of stellar gatherings. We have already described the subtle telescopic blur that is M1 (Fig. 1.11), the great Crab Nebula, but there is an additional sky blur to be seen, this one visible to the naked eye. This second object is the slight wisp of luminous cloud that constitutes the Pleiades galactic cluster—M45.

Additionally, there is the hidden galactic cluster, hidden at least to the naked eye, of the Hyades. Just like the best way to hide an individual tree is to place it within the spread of a great forest, so the stars that constitute the Hyades galactic cluster are scattered among both background and foreground stars. The Hyades hides in plain sight, its stars sprinkled across the region corresponding to the imagined head of Taurus and enveloping the bright star Aldebaran, although Alderbaran is not a member of the cluster. The Pleiades ride that part of the sky overlooking the Bull's imagined left foreleg shoulder. It looks like a luminous thumbprint on the celestial sphere, and even to the naked eye it is clear that the Pleiades are something entirely different (Fig. 1.20).

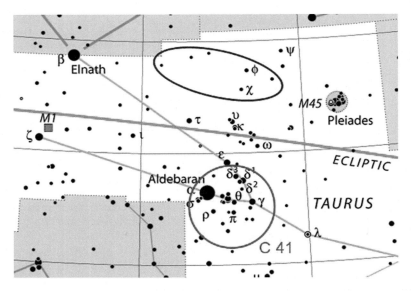

Fig. 1.20 Star map of Taurus showing the location of the Pleiades (*yellow disk*) and Hyades (*red circle*) galactic clusters—M45 and Caldwell 41, respectively. The *circles* indicate the approximate sky area over which cluster members are distributed. Also indicated to the *left* of the map is the position of M1—the Crab Nebula. The *blue ellipse* indicates the location of the nearest molecular cloud—TMC-1. Image adapted from www.iau.org/public/themes/constellations/

The Pleiades cluster has long been associated through Greek mythology with the seven divine daughters, or Pleiades, of the sea nymph Pleione, and this has resulted in the cluster's common name of the Seven Sisters. To the author's now-aged eyes the Pleiades looks like a small luminous cloud, about 1° wide, in the sky. Younger and keener eyes can reveal five or more stars in the cluster's glow. The question at this stage, however, is not how many stars are actually contained within the cluster but whether it is reasonable to assume that the clustering is real, and not just a chance, or random line-of-sight alignment of unrelated stars on the sky. This very question was, in fact, asked by Reverend John Michell[15] in a paper read to the Royal Society of London in May of 1767.

Mitchell set out to test the idea that stars might be "collected together in great numbers in some parts of space," and to do this he analyzed the probabilities of finding two stars a given distance apart on the sky. Comparing the observed closeness of stars in the Pleiades cluster with that which would be expected if all stars were distributed at random in the sky, Mitchell argued that the odds of a random grouping were "near 500,000 to 1." On this basis it would be highly unlikely that the stars seen in the Pleiades were simply the result of a chance grouping. If the grouping of stars within the Pleiades was not due to random chance, Mitchell noted, it was possible that they were brought together as a "consequence of some general law [such as gravity]." Mitchell was partially correct in his conclusion.

At this stage it is probably worth re-working Mitchell's analysis in modern form, and to this end we can use our star count data and apply a so-called Poisson statistics test. Named after the French mathematician Siméon Poisson, who developed the test in 1837, the Poisson test $P(r, \mu)$ is a measure of the probability that r stars are actually recorded in a given area on the sky when μ is the number of stars that would be expected to fall in the same given sky area purely by chance. Our guide to the following calculation will be Fig. 1.20, which reveals that the angular diameters φ for the Pleiades and Hyades are 2° and 6.7°, respectively. Again, from Fig. 1.20, we can count the number of stars brighter than magnitude +5, our comparative magnitude, for the Pleiades and Hyades regions, and the numbers are $r = 10$ and 15 respectively. Accordingly [10], the probability that the observed star groupings that constitute the Pleiades and Hyades have come about purely by random chance are of order 1 in 23.5 million and 1 in 400, respectively.[16]

Although we have more faith in our modern-day numbers, the conclusions that can be drawn from the above analysis is that Mitchell was correct in his basic idea that some regions of space do contain more, indeed, many more stars than would be expected purely by chance, and this suggests that such clustering is a real phenomenon. Indeed, the Hyades and the Pleiades are the two closest galactic clusters to the Solar System at the present epoch. The Hyades is located some 150 light years from the Solar System, with the Pleiades being some 410 light years distant.

---

[15] Mitchell is one of those greatly under-appreciated pioneers of modern science. He not only speculated upon the possibility of some cosmic objects having an escape velocity greater than the speed of light, a black hole in modern vernacular, he also developed the experimental technique behind the famed Cavendish experiment to determine the mean density of Earth.

[16] In terms of other probabilities, our analysis results are comparable to the 1 in 20 million odds of an individual eventually being canonized, and the 1 in 500 chance of being born with an extra finger or toe.

By combining large amounts of data relating to the distances and angular sizes of galactic clusters it turns out that they can be calibrated as standard rulers. In effect, the gravitational tides induced by the stars in the galactic disk limit the growth of clusters to about 40 light years across. Knowledge of this result, as we shall see in Chap. 3, was of great importance to Robert Trumpler in his 1930s investigation of the distribution of interstellar dust. Careful calibration of distance measures to the Hyades cluster[17] has resulted in it becoming an important standard for testing models of stellar evolution, and for representing the first rung on the cosmic distance ladder with respect to the so-called main sequence fitting method of distance evaluation.

Star clusters, as the detailed analysis given above implies, are important. In contrast to Mitchell's interpretation, that they are stars brought together by gravitational clustering, it is now known that clusters are really mass stellar birthings. The stars in a cluster form close together, all therefore having roughly the same age, and then gradually disperse over time into the background of stars in the galactic disk. Accordingly, as will be seen in later chapters, galactic clusters provide highly important information about the process of molecular cloud fragmentation, the process of star formation, as well as the aging effects of stellar evolution.

The Spectra of William Huggins

To the present-day tourist, the district of Tulse Hill, in the southern London borough of Lambeth, is not a place to inspire either tranquility or thoughts of scientific revolution. It is a bustling, noisy place. Some 150 years ago, however, in the quiet suburban surroundings of Upper Tulse Hill,[18] it was the center of astronomical research and innovation. Indeed, number 90 was the home of William Huggins, pioneer of the spectral analysis of starlight (Fig. 1.21).

Largely self-taught, Huggins spent the first 30 years of his life working in the linen and silk trade. Selling the family business in 1855, however, he turned his full-time attention to astronomy, building a private observatory in the garden of his new Tulse Hill home. His initial attention was directed towards the timing of occultations and the study of planetary markings. This all changed, however, in the early 1860s, when he heard of the pioneering spectroscopic work being performed by Gustav Kirchoff and Robert Bunsen in Heidelberg, Germany (see Chap. 2). Equipping his observatory with the appropriate experimental equipment Huggins, along with chemist and friend William Miller (who was in fact Huggins's neighbor), he set out to measure the spectra of the heavens.

The first two reports on their early findings appeared in back to back issues of the *Philosophical Transactions of the Royal Society* in 1864. The first communication considered observations relating to several bright stars (including Betelgeuse, Aldebaran, Sirius, Vega, Capella and Arcturus), and these all showed a broadly continuous spectrum crossed

---

[17] The Hyades cluster is so close to the Sun, in fact, that its distance can be determined through the observed proper motion of its members—by the so-called moving cluster method. Direct parallax distances have also been determined for individual stars in the cluster with the Hubble Space Telescope.

[18] Huggins's home at 90 Upper Tulse Hill no longer stands, the building being demolished, along with the homes of its neighbors, to make way for the modern housing estate of Vibart Gardens, Brixton.

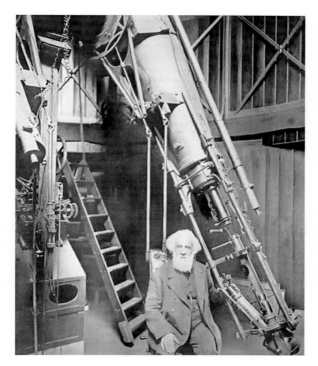

Fig. 1.21   William Huggins at his telescope in Upper Tulse Hill, London

by numerous absorption lines (see Chap. 2 for details). Indeed, the authors identified absorption lines due to sodium, magnesium, hydrogen, calcium and iron in the spectra of these stars. Speculating on the consequences of their findings, Huggins and Miller noted that the similarity between the spectra of the Sun and those of the stars that they observed indicated a commonality of structure and material composition. Additionally, they speculated that "the brighter stars are, like our sun, upholding and energizing centers of systems of worlds adapted to be the abode of living beings." In many ways, modern astrophysics was born with the publication by Huggins and Miller—stars becoming very much more than simple objects with just a position and brightness to measure.

The second of the papers published in the *Philosophical Transaction* by Huggins and Miller was the truly revolutionary one. In this work they focused "On the Spectra of Some of the Nebulae," and it is within this document that the diffuse nebulae are shown to be truly gaseous in composition. The first object that Huggins and Miller turned their spectroscope toward, on August 29, 1864, was NGC 6543, the Cat's Eye Nebula. This beautiful planetary nebula (Fig. 1.22) revealed not a continuous spectrum but a series of monochromatic emission lines. According to Kirchoff's laws (see Chap. 2), therefore, it could only be a hot gas. So shocked by what they saw, Huggins and Miller noted in their paper that at first they "suspected some derangement of the instrument had taken place." Eventually, however, they identified emission lines that they attributed to hydrogen and nitrogen.

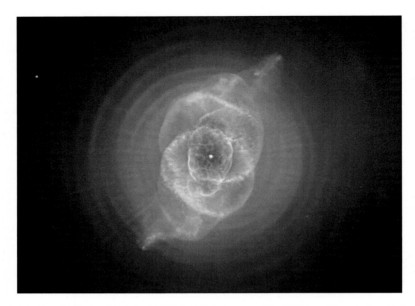

Fig. 1.22   The Cat's Eye Nebula (NGC 6543). Discovered by William Herschel on February 15, 1786, this planetary nebula is estimated to be about 3000 light years away. The nebula surrounds a massive binary star system composed of an O-type spectral star and a Wolf-Rayet star. It sits within a much larger but fainter halo structure of material that has been ejected from the central stars over the past 70,000 years. Image courtesy of NASA

In total eight nebulae were described by Huggins and Miller in their 1864 paper. Most were planetary nebula, including the (now) well known Ring Nebula (M57) and the Dumbbell Nebula (M27). Importantly, Huggins and Miller also examined a bright globular cluster in the constellation of Hercules (M92—recall Fig. 1.1c), and the Andromeda "nebula" (M31). Both objects showed spectra similar to that of a faint star, and there was no trace of any emission lines, which would have betrayed the presence of a hot gaseous component.

Here then was a further distinction between objects. The planetary nebulae were completely different in appearance, make-up and preferential sky location to the more common spiral nebulae—the term spiral nebulae having been introduced for objects such as M31 by Lord Rosse in 1845 (see Chap. 3). As Huggins and Miller put it, diffuse nebulae "can no longer be regarded as aggregations of suns after the order to which our sun and the fixed stars belong."

In spite of their momentous discovery, there was a problem associated with the spectra observed by Huggins and Miller. Indeed, they observed several green emission lines that had no known laboratory counterpart. Could it be, Huggins and Miller speculated, that the matter within gaseous nebulae is composed of a new kind of substance not known on Earth? It was to be a further 62 years, as we shall see, before the answer to this question was obtained.

Just 5 months after the reading of their first two papers at the Royal Society, a third paper, this time by Huggins alone, was made ready for publication. Given the title "On the Spectrum of the Great Nebula in the Sword-handle of Orion," this work set out to describe a "crucial test."

The test related to the resolution of diffuse nebulae, like that in Orion, into stars. Indeed Lord Rosse, using his great Leviathan telescope (72-inch diameter mirror), had argued in 1845 that the nebulae could be resolved into stars. Huggins spectroscopic analysis, however, clearly showed that Lord Rosse had been mistaken. The Orion Nebula showed the same three distinct emission lines observed in planetary nebulae, with no trace of the continuous spectrum expected for stars. It was indisputably a massive cloud of hot gas.

Finally, that vexing question that had so troubled both William and John Herschel was at least partially settled; the diffuse nebulae were revealed to be huge masses of glowing gas. Indeed, how the tables had turned. Just one decade before Huggins and Miller announced their results, preeminent physicist Sir David Brewster, in his 1854 publication *More Worlds Than Ours*, had confidently predicted, in support of John Herschel's position, that eventually all nebulae would be resolved into distant star clusters [11].

Although Huggins and Miller resolved the issue of diffuse nebula, there was still the problem of the spiral nebula to settle. Why were they distributed differently from the stars, and were they local or distant structures? Additionally, there was also the mystery of the nebulium lines to be dealt with.

Fortune smiled once again on Huggins and Miller in 1866 when a nova, literally a new star, appeared in the constellation of Corona Borealis.[19] Turning the spectroscope on this new stellar object, "the light from this new star formed a spectrum unlike that of any celestial body we have hitherto examined." Indeed, the nova showed a typical stellar absorption line spectrum, but superimposed on this was a set of bright emission lines. There was only one interpretation of the spectrum that could be made, and as noted by Huggins and Miller, "in consequence of some vast convulsion taking place in this object [the central star], large quantities of gas have been evolved from it." [12]

While Miller was already a Fellow of the Royal Society (indeed, he was treasurer from 1861 to 1900), the remarkable contributions to astronomy by Huggins were recognized by his election to that august body in 1865. The society awarded its Gold Medal to Huggins and Miller in 1867, and Huggins was further awarded the society's highest honor, the Copley Medal, in 1898. In 1869 Huggins pioneered the use of Doppler shift measurements, as first described by Christian Doppler in 1842, as a means as means of determining the radial velocities of stars [13].

In continued recognition of his important contributions to spectroscopy, the Royal Society re-equipped Huggins's observatory in 1871. Then, at the tender age of 51, Huggins married Margaret Murray, and the two embarked upon a highly successful partnership concerning the study of high-dispersion spectra.

Indeed, it appears to have been Margaret that first endorsed[20] the term nebulium, in 1898, to describe the supposed element responsible for the anomalous green emission lines seen in the spectrum of planetary nebulae. These emission lines appeared at wavelengths that had no known laboratory analog, a result that implied either a totally new element, not found on Earth, existed in space, or that the atoms responsible for their pro-

---

[19] This was the first outburst of the recurrent nova T Corona Borealis. A second nova outburst from the system was recorded in 1946 [12].

[20] In a delightful, but short communication to the *Astrophysical Journal* (volume 8, 1898) Margaret additionally suggested coronium, nephium and asterium as possible names.

duction were in a state not easily reproduced under laboratory conditions. Indeed, it turns out that the latter condition applied, but the understanding of atomic structure in terms of quantum mechanical principles had first to be developed.

Using both new experimental and theoretical procedures it was Ira Brown, at the California Institute of Technology in Pasadena, in 1927 who first demonstrated that the nebulium emission lines were in fact due to so-called forbidden line transitions in ionized oxygen and nitrogen. Such lines are only produced under extremely rarefied conditions, where the time interval between atomic interactions is very long.

The View Going Forward

Regions of lucid matter taking forms,
Brushes of fire, hazy gleams,
Clusters and beds of worlds, and bee-like swarms
Of Suns, and starry streams

—A. L. Tennyson

The reverend Hector Macpherson had a life-long interest in astronomy, and he wrote his first popular book on the subject, *Astronomers of Today and Their Works*, in 1905 when he was just 17 years old. Indeed, he wrote widely on astronomy and was well known for his public lectures. In 1911 he published his fourth book, beautifully entitled *The Romance of Modern Astronomy, Describing in Simple but Exact Language the Wonders of the Heavens*. It is a lovely book to read even today, although clearly many details have changed over the past 105 years. The *Romance* is over 330 pages in length, but Chapter 22, concerning the nebulae, is just five pages long. It is the shortest chapter in the entire text. For all this, however, the chapter begins, "Perhaps the most remarkable objects in the heavens are the hazy celestial clouds known as nebulae, or, as they have been picturesquely called, the fire mist." What are these nebulae, Macpherson questioned, responding that they can be divided into four main groups:

1. The irregular or extended nebulae (see e.g., Figs. 1.1a, and 1.4)
2. The planetary nebulae (see e.g., Figs. 1.6 and 1.8)
3. Dark nebulae (see e.g., Fig. 1.5)
4. The spiral nebulae (see e.g., Fig. 1.1b).

Nebulae in groups 1 and 2, Macpherson explains, are revealed through the spectroscope to be truly gaseous. These nebulae are described as belonging to the stellar system and are additionally concentrated towards the galactic plane. Planetary nebulae are described as being the "wrecks of ancient novae." The dark clouds of group 3 are somewhat mysterious and are attributed to "obscuring bodies, nearer to us than the distant stars." The spiral nebulae of group 4 are noted as being the most numerous in the heavens, with spectroscopic observations revealing them to be aggregated star systems. As to why such nebulae are seen predominantly in the directions towards the galactic poles, and as to whether they are independent island universes or local gas clouds in the process of generating new stars, or simply "the waste material of the universe," Macpherson notes, "[I]t cannot be said that astronomers are agreed."

In his short but precise Chapter 22, Macpherson provides us with a good overview of where the nebulae stand with respect to the deliberations of astronomers at the beginning of the twentieth century. The nebulae are now viewed as being entirely distinct from the open clusters that pervade the galactic plane and the globular clusters that predominate in the direction of Sagittarius. The full evolutionary and potential interconnected status of these objects was still unclear, however. Though planetary nebulae were associated with an end phase of stellar life, it was not then entirely clear where new stars might actually form—it not being clear at that time if the gaseous irregular and extended nebulae (group 1 objects) were associated with star formation or not. Additionally, the spiral nebulae were still a source of great confusion and diverse conjecture.

The continued story of the nebulae and their place in the star formation narrative will be pursued separately over the next four chapters. In Chap. 2 we shall begin to unravel the story behind the diffuse and gaseous nebula. Their status as star formation factories, embedded within giant molecular cloud structures, will be discussed in Chap. 4. In Chap. 3 we shall turn to the story relating to the dark and spiral nebulae and discuss what they tell us about our galaxy and the greater universe.

The place of planetary nebulae in stellar evolution, and their influence on the makeup of the interstellar medium, will be examined in Chap. 5. Likewise, the place of M1, the Crab Nebula, and similar such supernovae remnants, will be put into perspective in Chap. 5. To Macpherson and other astronomers at the turn of the twentieth century, M1 was just another group 1 extended nebula. Today the progenitor supernovae that are responsible for the production of such nebula are recognized as being the major movers and shakers of the interstellar medium. Indeed, they are fundamental triggers of star formation, and the major engines of chemical change in the universe.

REFERENCES

1. Halley investigated the possibility that the observed magnetic field variations across the Earth's surface might be mapped-out and thereby used as a means of determining longitude at sea. Unfortunately, as Halley discovered, the magnetic field variations were highly variable. Halley suggested in a paper read before the Royal Society of London in 1692 that the anomalous variations in the geomagnetic field could be explained if the Earth was composed of a series of concentric shells. Each shell was envisioned as having its own magnetic field and independent rotation. The magnetic field at the Earth's surface was then expressed as a combination of the time and position varying magnetic fields associated with the rotating shells.

2. If all stars had the same luminosity (L), then their distances (d) could be gauged directly from their observed brightness – or more precisely their measured energy flux (F) at the Earth's surface. The three quantities being related through an inverse square law in the distance, such that the measured flux $F = L / (4 \pi d^2)$ - see Appendix I in this book.

3. There is a long history of trying to solve Olbers Paradox, even before Heinrch Olbers brought it to popular attention in the 1820s. At issue is the contention that if in every line of sight direction that one could possible see on the sky there was a star, then the entire sky should be aglow with light. The paradox is to explain why this is observably not the case. The answer to this paradox entails the fact that the universe is not infinite

in extent or age, and that the universe is additionally expanding – the consequence of the latter condition being that as spacetime expands, so light is redshifted to longer and longer wavelengths. It is this latter effect, specifically, that accounts for the fact that at the present epoch the light produced during the Big Bang moment of creation of the universe is observed in the microwave part of the electromagnetic spectrum – that is as the cosmic microwave background radiation. As the universe continues to expand, so the cosmic background radiation will peak at successively longer and longer wavelengths.

4. Halley deduced the fact that the bright stars Arcturus, Aldebran and Sirius had shifted relative to the companion constellation stars by comparing contemporary positional data against that provided in several historical star catalogs.

5. The multiverse is a hypothetical set of infinitely many universes. The origins of this concept lie within the currently favored philosophical formulations of quantum mechanics and cosmology, and they attempt to explain the many fine-tuning issues that seem to apply to the observable universe. For a highly readable discussion of these topics see the recent book by Max Tegmark, *Our Mathematical Universe: my quest for the ultimate nature of reality* (Alfred A. Knopf, New York, 2014).

6. From the translation of Lacaille's notes – see the (highly recommended) Students for the Exploration and Development of Space (SEDS) web page: http://messier.seds.org/xtra/history/lac1755.html.

7. Messier's first 1774 catalog contained 45 objects. By 1781 the number of objects cataloged by Messier had increased to 80, and by 1784 the number had risen to 102. Additional objects have been added to Messier's catalog over the years, the last one being the dwarf elliptical galaxy M 110 (a satellite galaxy to M 31) by Kenneth Jones in 1967. Some of the objects in Messier's catalog might reasonably be excluded as being neither a physical grouping of stars (a galactic or globular cluster) nor a true nebulous cloud – M 24, M 40, M 73 are really just star asterisms. There is additionally, no present-day entry for object M 102, which is thought to have been a duplicate entry for M 101.

8. Key to the expanding nebula method working is the calculation of the actual expansion distance D in time interval T: this, in fact, is simply $D = T\,V\!exp$, where V_{exp} is determined by the Doppler shift method (see note 13) With D determined the distance to the nebula will be $d = D\,/\,\varphi$ (radians), where φ is the observed angular expansion, expressed in radians, of the nebula in time T.

9. Taking the number of stars per unit volume to be constant, the number of stars counted in two surveys, one out to a distance R_1 and the other out to a distance $R_2 > R_1$, will be $N_1\,/\,N_2 = (R_1\,/\,R_2)^3$. Since, additionally, the brightness of a star falls off as the inverse distance squared, we can relate the brightness of the faintest stars observed in each survey as $B_1\,/\,B_2 = (R_2\,/\,R_1)^2$ where B_1 is the brightness of the faintest star seen in the survey out to a distance of R_1, and B_2 is the brightness of the faintest star seen in the survey out to a distance of R_2. We now eliminate R_1 and R_2 from these two expressions and find $B_1\,/\,B_2 = (N_2\,/\,N_1)^{2/3}$. At this stage it will be convenient (from a modern perspective) to introduce the idea of the magnitude scale of stellar brightness. The magnitude scale (see Appendix I) is based upon the measured brightness (more specifically, the measured energy flux) at the observer's telescope. If the measured energy flux

from a specific star is f^* (joules per square meter per second), then the magnitude m^* is expressed as $m^* = -2.5 \text{ Log } (f^*)$. This formulation and conversion was introduced by British astronomer Norman Pogson in the mid-1850s. If we now take the logarithm of the ratio B_1 / B_2 we obtain the result that $\text{Log } (B_1 / B_2) = (2 / 3) \log (N_2 / N_1)$. Now in accordance to the manner in which Pogson's scheme is set up, the magnitude difference $(m_1 - m_2)$ between two stars of brightness B_1 and B_2 is, $m_1 - m_2 = -2.5 \log (B_1 / B_2)$, and so $m_1 - m_2 = -2.5 \log (B_1 / B_2) = -2.5 (2/3) \log (N_2 / N_1)$. Finally, if we 'fix' our surveys so that the faintest stars observed in the survey to distance R_2 is one magnitude fainter than the faintest stars observed in the survey out to R_1, then $\log (N_2 / N_1) = 1 / (2.5 \times 2 / 3) = 0.6$, which can be expressed yet more simply as $(N_2 / N_1) = 10^{0.6} = 3.981$. That is, what our initial assumptions of uniform star distribution and constant star luminosity imply is that if we conduct two star-count surveys, one to a limiting magnitude of m and the other to a limiting magnitude of $m + 1$, then $N_{m+1} / N_m \approx 4$)

10. From Table 1.1, we see that our all-sky count for stars brighter than +5 is $N_5 = 2{,}724$, and the final step is to calculate the μ parameter, which is given as the ratio of N_5 / N_A, where N_A corresponds to the total number of cluster areas $A_C = \pi (\varphi / 2)^2$ that would be required to cover the entire sky: $N_A = 41{,}253 / A_C$, where the 41,253 number corresponds to the number of square degrees over the entire sky. The probability that the star groupings in these two cluster came about purely by random can now be evaluated as $P = N_A P(r, \mu) \times 100$. The multiple by N_A in the probability comes about because we do not specifically care which area A_C we are looking at on the sky. For counting experiments such as we are considering, where events occur at a random rate μ, the probability of recording r events in a particular trial is give by the Poisson distribution: $P(r, \mu) = (\mu^r / r!)e^{-\mu}$, where $r!$ is the r-factorial product $r(r-1)(r-2)\ldots.3.2.1$.

11. Sir David Brewster (1781 – 1868) specialized in optics, and is credited with the invention of the binocular camera, the stereoscope, various polarimeters and the kaleidoscope. He was one of the founders of the British Association for the Advancement of Science, a long time editor of the Edinburgh Philosophical Journal, and he edited the first 16 volumes of the Edinburgh Encyclopedia from 1824 to 1830. The proof that Brewster offered in his *More Worlds than One* (published in 1854), is a classic example of incorrect but precise logic. In short he offered an inductive proof for the resolvability of all nebulae into star clusters. He argued that each time a new and larger telescope is constructed, so more and more nebulae are resolved in to stars. Taking the Orion nebula as an example, Brewster noted that Galileo in the early 1600s failed to see it at all with his telescope, William Herschel in the mid-1700s saw it as a diffuse nebula, but Lord Rosse (erroneously as we now know) in the mid-1800s resolved it into stars. Accordingly, so argued Brewster by inductive logic, those nebulae seen by Rosse must ultimately be resolved into star clusters when new and larger telescopes are brought to bear on them. The logic seems impeccable, but it is entirely based upon the unarticulated premise that all nebulae must be star clusters.

12. The star T Corona Borealis is a member of that group of variable outburst stars known as cataclysmic variables. Such binary systems are comprised of two stars that are sufficiently close to each other that mass transfer from one star to the other can take place. If the accreting star is an evolved white dwarf object an accretion disk can form around it, and this disk may, at times, become unstable. A disk instability episode can

result in the rapid and massive accumulation of matter on the surface of the white dwarf, and this in turn can result in the sudden onset of fusion reactions, triggering a nova outburst. During a nova outburst the brightness of the system increases dramatically, and much of the material that accumulated on the surface of the white dwarf will be ejected into the surrounding interstellar medium.

13. The Doppler Effect is concerned with the apparent change in the observed wavelength of an emission or absorption line feature due to the relative motion between the star and the observer. The difference between the observed wavelength λ_{obs} and the expected wavelength λ if the star and observer were at rest is related to the radial velocity V_R as: $(\lambda_{obs} - \lambda) / \lambda = V_R / c$, where c is the speed of light. Austrian physicist Christian Doppler first described the basis of the phenomena named after him in 1842, although in the original publication he incorrectly attributed the different colors observed for the stars as being due to their relative motions towards or away from the Earth – these color difference are now known to be associated with the different temperatures of stars (a consequence, in part, explained by Wien's law – see chapter 5 [2]).

2

In the Eye of the Beholder

Certain it is, that although our conclusions may be incorrect,
Our judgment erroneous, the laws of nature and the signs afforded
To man are invariably true. Accurate interpretation is the real deficiency.

—Rear Admiral Fitz Roy, *The Weather Book* (1863)

Let There Be Light

The eye, human or otherwise, is one of the most wonderful creations of nature. Through the steady searching of accumulated adaptation, working through the random shuffle of mutation, and the rigid testing of natural selection, evolution has found that special sweet spot of design that has enabled life to see.

The eye has evolved independently many times throughout the history of life on Earth, and it has followed many pathways of design. Some eyes are compound, some use lenses, some use mirrors, some work like pinhole cameras. Many eyes have evolved receptors that are sensitive to the visual wavelengths of light; other eyes are more sensitive to the shorter wavelengths of ultraviolet light, and yet others to the longer wavelengths of the infrared. For all the variety in light-manipulation design and wavelength sensitivity, however, all eyes are doing one really important thing—they are looking. The eye enables knowledge of the world to be ascertained, exploiting the way in which objects react with, that is reflect, refract and/or absorb, electromagnetic radiation emitted into space by the Sun. Evolution found the design for the eye, but it was the Sun that (at least partly) constrained its direction of exploration and search.

In 1964, Marshall McLuhan famously coined the phrase "the medium is the message," by which he intended to convey the idea that the form of the medium influences the way in which the message is perceived. For light, or more correctly electromagnetic radiation, the reverse holds true. The message is entirely encoded in the medium, and the medium is composed of a composite variation in the local magnetic and electric fields. For the

© Springer International Publishing AG 2017
M. Beech, *The Pillars of Creation*, Springer Praxis Books, DOI 10.1007/978-3-319-48775-5_2

astronomer, the art of making progress, that is, learning more about an object or phenomena, is all about finding the hidden message that has been interwoven within the light emitted by stars.

It was the great physicist James Clerk Maxwell that gave us the modern theory of light in the mid-1860s. His idea was bold, highly mathematical, and largely ignored—at least initially. The question that Maxwell asked, and then successfully answered, was, when an oscillation or disturbance is set up in a magnetic field, how does that disturbance evolve over space and with time? This hardly seems like a question that might lead to an understanding of light, but the key point of Maxwell's analysis was that any disturbance that might be introduced into a magnetic field will result in the generation of a wave, a somewhat complex wave involving twin oscillating electric and magnetic field components, but a wave nonetheless, that will move through the vacuum of space at the speed of light.

It was the speed of light part to his wave equation that set his mind running. If an electromagnetic wave moves through space at the speed of light, he reasoned, so the phenomenon of light itself must surely be an electromagnetic wave. The logic was inescapable: if A equals B, then B must equal A. We need not worry about the full details of Maxwell's electromagnetic radiation theory here, but at this stage some very basic ideas should be introduced [1]. Specifically, we should address the question of what is a wave, and then consider how one type of wave can be distinguished from another. These are in fact old questions, and they predate by centuries Maxwell's theory of light.

So, what is a wave? There is the immediate minds-eye response to this question. A wave is something like the curve of water rolling onto a beach. Yes, waves are encountered at the seaside, but they are actually very complex waves involving aspects of surface tension, sloping shorelines, wind direction, and gravity. A simpler picture of a wave might be that of a ripple moving along the surface of millpond or canal waterway, but even these waves are quite complex in their (constant soliton) form.

Perhaps we should move away from water waves. After all, waves do come in many shapes and arise under many different circumstances. There are earthquake waves through rock, there are traffic waves of cars upon city streets, there are sound wave pulses moving through the air, and there are electrons flowing like on-off waves through a computer device. Waves are truly present everywhere, and they have many different characteristics, but the one thing that they all do is carry information, that is, energy, from one location to another. Indeed, a wave is really just a disturbance that propagates through some specific medium (be it water, air, or rock).

A wave can be irregular in its profile, or it can be like a perfect sine wave; it can vary in shape with time and/or distance traveled, or it can remain forever constant. Indeed, light has the remarkable property that if left alone it will propagate forever, never changing its profile no matter how far it travels or for how long. In principle, light is eternal and immortal. So, if a wave is a disturbance that propagates through a medium, how is it that light can move through a vacuum that contains no medium whatsoever?

The answer to this takes us back to Maxwell. Although a true vacuum has no material present within it, no atoms, no electrons, and no dust particles, to support a pressure wave (such as sound), it can still support a magnetic and electric field, and it is this property of space that allows electromagnetic radiation to move through space. Additionally, an electromagnetic wave is self-generating as time goes by, and this, in fact, is the reason why it is a two-component wave, with magnetic wave and electric field components moving in synchronization at right angles to each other. The one, as Maxwell revealed, lives off the other.

A time-varying magnetic field generates a time-varying electric field, and a time-varying electric field generates a time-varying magnetic field, and so on *ad infinitum*. At this stage we simply have the notion that waves describe a process by which a disturbance propagates from one region to another, through some intervening medium, whether that medium is solid, liquid, gas, or a vacuum. What is being propagated is information in the form of energy (more on this shortly), and it is this energy information that moves from one region of the medium to another at a specific speed c. For a sound wave moving through dry air, $c = 340$ m/s, while for electromagnetic radiation, moving through a vacuum, $c = 299,792,458$ m/s, and the latter is the fundamental constant corresponding to the speed of light.

All electromagnetic waves propagate through space at the same speed, but the way in which they interact with matter and the way in which they can be detected vary according to their associated frequency (f) and wavelength (λ) (Fig. 2.1). The wavelength is the spatial distance over which the wave profile starts to repeat itself, while the frequency is related to the number of wavelength units that are repeated per second.

Straightforward dimensional analysis indicates that the result of multiplying the wavelength of a wave by its frequency must be equal to its speed: $\lambda f = c$. Since all electromagnetic waves travel at the speed of light, the longest wavelength waves must have a low frequency and the shortest wavelength waves must have a high frequency. It is this specific wavelength and frequency association that dictates the way in which an electromagnetic wave will interact with matter, and it will also determine the way in which the wave might be used.

Light is all around us, provided in great abundance by the Sun, but the human eye is only sensitive to electromagnetic radiation in the wavelength region running from 380 nm, corresponding to blue light, to 900 nm, which corresponds to red light. That light, that is, electromagnetic radiation, can have wavelengths much shorter and much longer than those perceived by the human eye was, in fact, demonstrated experimentally some 60 years

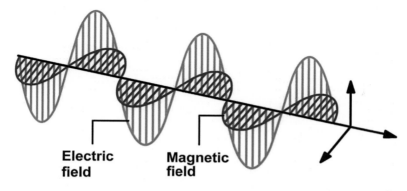

Electric field   Magnetic field

Fig. 2.1   An electromagnetic wave. The electric and the magnetic field variations occur at 90° to each other, and the amplitude of the field strengths trace out a sine wave curve. Image courtesy of NOAA

before Maxwell outlined his famous theory. At the time that these early experiments were first performed, however, it was not realized that they had anything in common, or indeed that they had anything to do with electric and magnetic fields.

The first indication that there might be more to the property of sunlight than its power to illuminate the world around us was first revealed by astronomer William Herschel in 1800. In an experiment remarkable for its ingenuity and shear simplicity, Herschel set out to measure the temperature of the different colors of sunlight. He used a large prism to spread out the color spectrum, blue to red, of sunlight and then simply set a thermometer at different locations to determine the temperature (Fig. 2.2). Serendipitously, within this

Fig. 2.2 An artist's schematic interpretation of the sunlight prism experiments conducted by William Herschel in 1800. Image courtesy of IPAC and Caltech

experiment, Herschel discovered that the highest temperature was recorded when the thermometer was placed in a region beyond that of the red light rays, in a region where the eye could see no color at all. This heat ray form of light is now known as infrared radiation.

In 1801 German chemist Johann Ritter, inspired by Herschel's heat ray result, set out to see if there were any detectable rays in the region beyond that of blue light. Once again Ritter used a prism to separate out the colors of sunlight, but this time, along the resultant spectrum he set out grains of silver chloride crystals. Ritter knew that silver chloride crystals reacted more strongly with light as the color shifted from red to blue. Remarkably, however, he found that the silver chloride grains reacted most strongly when placed in a region situated beyond that of blue light, again in a zone where the eye could see no illumination at all. To describe his newly discovered effect, Ritter introduced the term oxidizing rays, but we now call such rays ultraviolet radiation.

New forms of light at wavelengths both longer and shorter than those detected by Herschel and Ritter were discovered towards the close of the nineteenth century. German physicist Heinrich Hertz set out initially to prove Maxwell's theory wrong, but in 1887 he developed and performed an experiment with oscillating coils and a spark gap resonator that not only proved Maxwell's electromagnetic wave theory but identified a new form of radiation with a wavelength that placed it well beyond the heat rays of Herschel. These new rays form the radio wave part of the electromagnetic spectrum, and while Hertz dismissed his discovery as being "of no use whatsoever," radio transmitters, as designed by Guglielmo Marconi and Nikola Tesla, were soon beating out the staccato rhythms of Morse code around the world, and eventually, by the second decade of the twentieth century, commercial radio stations came on air. X-rays, with shorter wavelengths than Ritter's ultraviolet radiation, were discovered serendipitously in 1895 by German physicist Wilhelm Röntgen.

Experimenting with the newly discovered cathode rays,[1] Röntgen chanced to notice that when his vacuum tube experiment was running a card that had been coated with barium platinocyanide began to glow. This fluorescent effect, Röntgen reasoned, indicated the emission of a new kind of ray, a ray invisible to the eye, and, as subsequent experiments showed, a ray with considerable penetrating power. These new X-rays (where the X was used to indicate something that was unknown) could even penetrate skin and muscle, to reveal the bones hidden beneath. Gamma rays, having even smaller wavelengths than Röntgen's X-rays, were discovered by the French physicist Paul Villard in 1900 while performing experiments on radium salts.

Over the span of exactly 100 years, the domain of electromagnetic radiation had been mapped out by serendipity and direct investigation. From the longest wavelength radio waves to the shortest wavelength gamma rays, the extent of the electromagnetic spectrum began to unfold (Fig. 2.3), and the various utilitarian properties of its associated rays were developed. Our modern world of cell phones, television, and radio all exploit radio wave communication; microwaves (having a wavelength of a few cm) heat our food; firefighters use infrared cameras to seek out hot spots in a fire; ultraviolet light is used to sterilize the water that we drink; doctors use X-rays to identify internal ailments; and engineers use gamma rays to search for cracks and defects in massive concrete structures. Electromagnetic

---

[1] Although called rays, this phenomenon is actually related to a flow of charged particles.

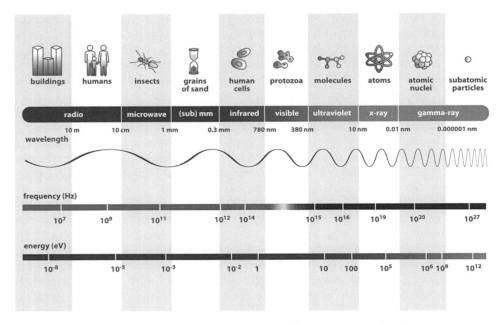

Fig. 2.3   The electromagnetic spectrum. Image courtesy of ESA

radiation is all around us, and its uses are many and profound. This same multitude of applications is also exploited by astronomers (as we shall see later) to reveal those features of the universe that are otherwise hidden and entirely invisible to our eye.

A Quantum Mechanical Aside

German physicist Max Planck introduced the concept of quantized energy to explain a phenomenon that had dogged classical physicists throughout the latter decades of the nineteenth century. The problem related to an experiment concerning the way in which hot objects radiate energy into space. Indeed, it had been observed that when the amount of energy radiated per second per square meter by a hot object is measured, the energy flux F_λ is found to not be constant with wavelength. For any given temperature there was a well-defined wavelength at which a maximum energy flux appeared, but either side of this maximum, of course, the energy was smaller and fell off towards zero at the longest and shortest wavelengths.

The variation of the energy flux with wavelength for such 'blackbody' radiators was initially a complete mystery, and it had no explanation within classical physics. Indeed, classical physics clearly and unequivocally predicted that the bulk of the energy should be radiated at short wavelengths, that is, in the ultraviolet part of the spectrum. The experimenters, in complete contrast, found the exact opposite to the theoretical prediction—the energy flux decreased rapidly at shorter wavelengths. Nature, it would seem, had out-smarted the theorist once again.

As a newly graduated theorist, Planck decided to take on the problem, and he ultimately decided that the only way to make progress was to introduce a mathematical trick. The trick was, in effect, to argue that the radiated energy was supplied in small fundamental packets, or quantum units, with each unit being supplied at a level according to the product hf, where h is now known as Planck's constant and f is the frequency of radiation. This mathematical trick, which made the theory fit with the observational data, was the beginning of quantum mechanics, and in Planck's words was introduced "as an act of desperation." It was a desperate measure because, at that time, he had no justification for explaining why energy was distributed in little packets rather than according to a continuum (as demanded by classical physics).

Planck introduced his new results to the assembled members of the German Physical Society in Berlin, on October 19, 1900. The audience was not impressed, because there was no justification for the mathematical trick, and there was no expedient for restricting energy to be supplied only in small quanta. History reveals, however, that Planck's little trick was not only mathematically sound but physically justifiable as well. Indeed, quantum mechanics is now one of the major pillars of modern physics.

As a result of Planck's initial idea and the gradual development of quantum mechanical rules we now have to live with a strange duality in the way that light is to be envisioned. We can endow electromagnetic radiation with wavelike characteristics such as wavelength and frequency, but when we ask about the energy associated with an electromagnetic wave it is expressed in terms of a particle-like quantity, or energy quantum, $E = hf$. This wave-particle duality is directly observed in the sense that in some experiments electromagnetic radiation behaves as if it is a wave, while in others it behaves like a particle (or, to use the expression introduced by chemist Gilbert Lewis, like a photon) carrying a specific quantum of energy.

It has been said of the wave-particle duality phenomenon that it is the wave part that determines where the particle part goes, but it is the particle part that tells the wave part when to begin and when to stop. This is really only an approximate picture of what is actually going on—the devil, as always, is in the details—but it gets us to the core of the issue at hand, and it brings us to a few technical points that have a direct bearing on later discussion.

Unweaving Light

Is it a wave or is it a particle? Such was the question concerning the physical properties of light that engaged the minds of philosophers towards the close of the seventeenth century. And beyond the purely factual, the answer that one chose to believe had important ramifications. Indeed, the answer played into the politics of one of the greatest historical rivalries of science.

The taciturn but brilliant Isaac Newton preferred an explanation in terms of particles. The ever inventive and sociable underdog Robert Hooke preferred an explanation in terms of waves. It turns out they were both right in their basic concepts but both wrong with respect to the details. It was Newton, however, who provided the framework by which light could be analyzed. "…[I]n the beginning of the year 1666… I procured me a Triangular glass-Prisme, to try therewith the celebrated Phenomena of Colours…" So wrote Newton in a letter to the Royal Society in 1671.

Certainly there was nothing new in the starting premise, it being known long before that a triangular block of glass could produce a rainbow spray of colors. What was new, of course, is the genius that Newton was to bring to its use as an experimental device. Newton took the prism and then set about, in grand Baconian fashion, putting nature to the test. In his *experimentum crusis*, Newton specifically argued that sunlight could be split, with the aid of a triangular section of glass, into a continuous spectrum of colors—the colors ranging across the rainbow pattern from red through yellow to green, blue, and violet.

For Newton, however, the key outcome of his experiments was that light must be a "heterogeneous mixture of differently refrangible rays," and that each color has as a distinct and constant refrangibility in a given medium. In other words, Newton argued that the rays that characterize every color must pre-exist in white light. Opticians no longer use the word refrangible, but rather substitute the concept of refractive index as a measure by how much the path of a monochromatic light ray is changed when it passes from one medium to another. This result is encompassed within what is now known as Snell's law, so named after the Dutch astronomer Willebrord Snellius who first[2] derived the mathematical formalism behind the refraction of light in 1621.

For astronomers the utility of Newton's experiment was that it allowed for an understanding of starlight, although this crucial connection was not made until the early nineteenth century and the unexpected observations recorded by British chemist William Hyde Wollaston, in 1802, and German optician, Joseph von Fraunhofer, in 1812. Although Wollaston appears to have been the first person to study the spectrum of the Sun, and observed a continuous spectrum crossed by a number of dark lines, it was Fraunhofer who presented the first detailed analysis (Fig. 2.4). Indeed, Fraunhofer is credited with developing the first spectroscope, in which the light to be examined must first pass through a narrow slit before encountering the triangular prism of glass, where the resultant spectrum is analyzed in detail with a telescopic eyepiece (Fig. 2.5). It was in this manner that he discovered and described the celebrated anomaly of what are the Fraunhofer lines.

Fig. 2.4 A series of spectra showing the continuous color spectrum (*red to blue*) of the Sun along with several stars and a nebula, all of which are crossed by dark absorption lines. In order from the top are the spectra for the Sun; the *blue and red* components to the binary stars Albireo (β Cygni), Sirius, and Rasalgethi (α Herculis); and the recurrent nova T Corona Borealis. The last four spectra show the emission lines of hydrogen, nitrogen, and carbon associated with the Cat's Eye Nebula in the constellation of Draco. The image is from Henry Roscoe's *Spectrum Analysis* (first published 1878). The lettering for the strongest Fraunhofer lines is shown in the top spectrum

---

[2] The law of refraction has actually been derived independently many times throughout history. Before Snellius, the same basic relationship had been discussed by Thomas Harriot in 1602, and the geometry of the situation was outlined by Arabic scholar Ibn Sahi in 984. After Snellius, René Descartes famously re-derived the result in 1637.

S P E C T R U M

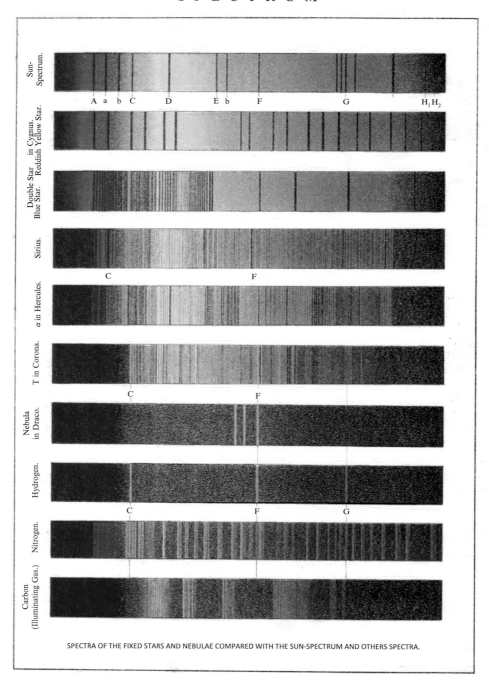

SPECTRA OF THE FIXED STARS AND NEBULAE COMPARED WITH THE SUN-SPECTRUM AND OTHERS SPECTRA.

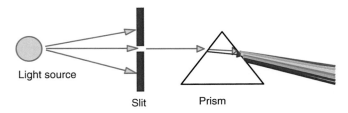

Fig. 2.5 The basic design principles of a prism spectroscope. Starlight, with the aid of a telescope, is first brought to a focus on the spectroscope slit. The exiting spray of light is then collimated and brought to one side of a triangular glass prism, where, upon passing through it, the light is split into its different wavelength-dependent color components. Finally the resultant spectrum, along with associated absorption lines, is examined in detail with a small viewing telescope. Image courtesy of the Australian Telescope National Facility

The origin of the dark lines observed to cross the continuous (rainbow color) spectrum of the Sun was initially a complete mystery, but it was soon realized that they must be related to the Sun's chemical composition. Indeed, by 1814 Fraunhofer had determined the precise wavelengths of some 350 lines, of which the most prominent were given letter designations (Fig. 2.4). The A-line, at a wavelength of $\lambda = 0.759$ μm (in the blue region of the continuous spectrum) is now known to be due to O_2 molecules in Earth's atmosphere. A pair of strong lines in the yellow part of the continuous spectrum was labeled D, and this corresponds to the element sodium. A strong line in the blue-green region of the spectrum was designated F, and this was found to be due to the element hydrogen. An additional set of closely spaced strong lines in the violet region of the spectrum were designated H and K, and these are due to singly ionized calcium Ca^+ atoms.

Some 25,000 Fraunhofer lines have now been recoded in the Sun's spectrum, which provide a detailed signature of the Sun's composition. By 1823 Fraunhofer had extended his spectral studies to bright stars and planets, but by 1826, at just 39 years of age, this dedicated pioneer was to die of tuberculosis.

The application of spectroscopy to laboratory studies, however, continued apace, and new practitioners such as David Brewster, William Fox Talbot, and John Herschel, during the 1840s, were able to develop both the instruments and the measuring techniques that would allow for the association of various spectral lines with specific chemical elements. Indeed, it was the double D line pair, as labeled by Fraunhofer, which were first identified and associated with the element sodium. There was a phenomenological puzzle, however, related to this identification. When sodium was sprinkled into a flame, and the spectrum observed, yes, the flame turned green, and yes, the D lines were present, but the spectroscope showed that the D lines in the flame were bright and distinct. The wavelengths of the D lines in the flame, however, were nonetheless identical to the wavelengths of the dark D lines observed in the Sun's spectrum. The question arose, therefore, as to what are the exact circumstances that determine whether a specific line will be seen in emission (that is, bright) or in absorption (that is, dark).

The answer to this question was effectively provided in 1859 by Robert Bunsen and Gustav Kirchoff. Indeed, it was these two practitioners who outlined in a research paper published in the *Philosophical Magazine* in 1861 the great possibility of using spectroscopy to study the very makeup and composition of the stars—opening up, thereby, "an entirely untrodden field stretching far beyond the limits of the earth, or even the Solar System."

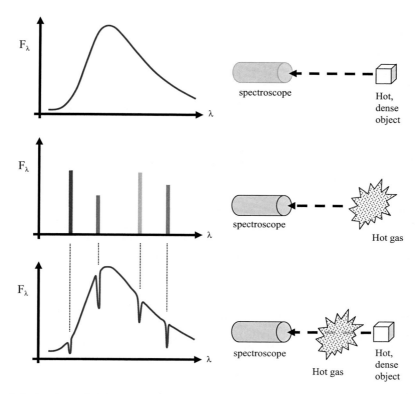

Fig. 2.6 A schematic illustration of Kirchoff's laws. *Top:* A continuous (blackbody) spectrum as emitted by a hot, dense liquid, solid, or gas. *Middle:* A discreet emission line spectrum as emitted by a hot gas. *Bottom:* An absorption line spectrum, as observed when a hot, dense object is seen through an intervening cloud of cooler gas. Note that the absorption lines fall at exactly the same wavelengths as the discrete emission lines

The critical experiment performed by Bunsen and Kirchoff in 1859 was to pass a beam of sunlight through a sodium flame. What they observed was that the D lines became even darker, and this they reasoned indicated that the hot sodium flame had both absorbing and emitting properties.

What are now known as Kirchoff's laws emerged from this pioneering study, and these laws state that only three kinds of spectra can be produced (Fig. 2.6). It transpires, according to Kirchoff's laws, that hot solids, liquids, and very dense gases will produce continuous spectra, with light (electromagnetic radiation) being emitted at all wavelengths. In contrast, a hot, low density gas will produce an emission line spectrum, with electromagnetic radiation being emitted at a number of discrete wavelengths—there being no continuum emission. An absorption line spectrum will be observed, however, when the continuous spectrum produced by a hot, dense solid, liquid, or gas is observed through a lower density, lower temperature gas situated along the line of sight. It is the intervening gas cloud that produces the absorption lines.

Returning to Fig. 2.4, we now see that stars produce absorption line spectra, while hot diffuse nebula (HII regions) are characterized by an emission line spectrum (as first observed by Huggins and Miller in 1864). That stars have an absorption line spectrum tells us that the density and temperature of the Sun must increase from the surface inwards. It is the deep, high density, very high temperature gas within the body of the star that generates the continuous spectrum. While it is the cooler, lower density gas in the star's outer layers that produces the absorption lines. Indeed, the absorption lines are produced within just the outer few hundred km of a star's photosphere. The quantum mechanical processes responsible for generating absorption and emission lines will be discussed in the next section of this chapter.

Although we shall leave the topic of stellar spectra here, history does record that spectroscopic studies have allowed for both an understanding of stellar atmospheres, and for the establishment of a spectral classification scheme by which similar stars can be recognized [2]. Additionally, one of the most important results from the analysis of stellar spectra is that the continuum component behaves in a manner close to that of a blackbody radiator. Indeed, the term blackbody (*schwarzen körper*) radiator was coined by Gustav Kirchoff in 1860, and he also initiated experiments to measure the energy flux F_λ, over a range of wavelengths, of such objects constructed in the laboratory. By definition a blackbody radiator is an object that absorbs all wavelengths of electromagnetic radiation that falls upon it. Hence, it is black because there is no reflection. For all this, however, the object must also re-radiate energy back into space or else its temperature would simply continue to increase as more and more radiation is absorbed.

Although a somewhat idealized object, a good approximation to a blackbody radiator can be realized by drilling a small hole into the casing of a constant temperature oven. Any light that chances to enter the hole will be absorbed within the oven's interior (hence the hole is black). Any light escaping through the hole, however, will emulate that of a blackbody radiator and will be entirely characterized in terms of the oven's interior temperature T. In the late 1870s such experimental blackbody (sometimes called cavity) radiators were studied in detail by such practitioners as John Tyndall in England and Pierre Dulong and Alexis Petit in France. Indeed, using the experimental data of these earlier researchers as his guide, Austrian physicist Josef Stefan deduced in 1879 a law relating the total energy flux F emitted (over all wavelengths of the electromagnetic spectrum) by a blackbody radiator of temperature T. This law, now known as the Stefan-Boltzmann law,[3] gives $F = \sigma T^4$, where σ is the Stefan-Boltzmann constant.

Using his newly derived law, Stefan, in his 1879 paper to the Vienna Academe of Science, derived the first good estimate for the temperature of the Sun, placing its surface temperature between 5580 and 5838 K (the modern-day value for the Sun's effective temperature is 5778 K). A second fundamental law relating to blackbody radiators was derived by German physicist Wilhelm Wien in 1893. Indeed, Wien's law relates the wavelength λ_{max} at which a blackbody emits its greatest energy flux to its temperature, with the product $\lambda_{max} T = $ constant.

---

[3] Ludwig Boltzmann provided a derivation of Stefan's empirical results, on theoretical grounds, in 1884.

Wien was awarded the 1911 Nobel Prize for Physics in recognition of this result, and he also provided astronomers with a rough-and-ready thermometer[4] by which the temperature of a star can be determined. Indeed, the temperature of a star can be estimated by finding the wavelength λ_{max} at which the greatest energy flux is recorded, and the color of a star will be governed by the location of λ_{max} in the visual part of the spectrum. A cool star has λ_{max} towards the red and infrared part of the spectrum, while a very hot star will have λ_{max} towards the blue and ultraviolet part of the spectrum.

Wien's law also provides a rough-and-ready guide to what sort of telescope and detector can be used to study various astrophysical phenomena. Low temperature phenomenon, such as isolated gas clouds in the interstellar medium or regions of early star formation, will best be studied with infrared and microwave telescopes. Very high temperature gas, such as that produced in a supernova explosion, or the coronal gas component (see Chap. 4) of the interstellar medium, will be best studied at ultraviolet and X-ray wavelengths. Intermediate temperature gas clouds, such as HII regions (see later), are best studied at optical wavelengths.

The full theoretical explanation concerning the behavior of blackbody radiators, and a full theoretical underpinning of the Stefan-Boltzmann and Wien's laws, as discussed earlier, was provided by Max Planck in 1900. A clear understanding of the emission line and absorption line processes, however, took a further quarter-century to fully evolve, but the intellectual pathway to understanding such lines was opened up by Planck's introduction of quantized energy.

The Color of Excitation

Before an understanding of emission and absorption lines was possible, an understanding of atomic structure first had to be developed, and especially an understanding of the ways in which electrons can interact with electromagnetic radiation. Indeed, the intellectual and experimental struggle to understand the atom was a mighty one, and many a famous physicist, chemist, and astronomer made their names by tackling the complex issues at hand [3]. Key to making progress was the development of quantum mechanics, and especially the idea of quantized energy.

Hard on the exploratory footsteps of Ernst Rutherford, who realized in 1911 that atoms must have a configuration that corresponds to a small central nucleus surrounded by an extended cloud of electrons, Dutch physicist Niels Bohr set out to apply the new quantization rules to its structure. The quantum mechanical underpinning of the Bohr atom, first introduced in 1913, required that the electrons only occupy orbitals (available space within the atom for an electron to reside) with very specific energies—the allowed energy levels being determined according to the species of atom. The lowest energy orbital, or ground

---

[4] Since stars are not perfect blackbody radiators astronomers don't actually use Wien's law to determine temperatures. Rather the characteristics of a star's absorption lines are analyzed to deduce an effective temperature. The effective temperature is the temperature of a blackbody radiator having the same total energy flux as the star being observed.

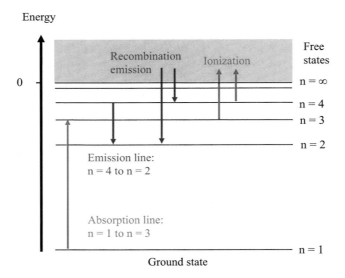

Fig. 2.7 A schematic energy level diagram showing a sample of allowed transitions. Electron transitions where the initial quantum number is smaller than the final quantum number (shown by *red arrows*) indicate absorption line transitions. Electron transitions where the initial quantum number is larger than the final quantum number (shown by *blue arrows*) indicate emission line transitions. Electron capture events produce recombination lines, while electrons absorbing more energy than the ionization limit will no longer be bound to the atom's nucleus

state, has a principle quantum number $n=1$, with the next allowed orbital, corresponding to the first excited state, having a principle quantum number of $n=2$, and so on all the way to the ionization limit $(n=\infty)$, for which orbital the electron essentially has zero energy and is no longer bound to the nucleus. Though an electron must reside in one or another of the allowed electron orbitals for the particular atom under consideration, it need not always stay in the same orbital. Indeed, electrons can move between allowed electron orbitals but only if they gain (that is, absorb) or expel (that is, emit) the exact amount of energy that distinguishes the initial and final orbitals.

For an absorption line transmission from the $n=1$ ground state to the $n=3$ orbital (Fig. 2.7), the electron has to absorb an amount of energy exactly equal to $E(3)-E(1)$, where $E(n)$ corresponds to the energy associated with the nth orbital. The electron can gain energy either through atom-on-atom collisions or by interacting with the surrounding radiation field. Indeed, the exact wavelength of the absorption line can be determined from Planck's energy formula: $E_{1\rightarrow3}=h\,(c/\lambda_{line})=E(3)-E(1)$.

The exact same principle applies to an emission line transmission, where the electron has to lose (that is, emit) energy in order to move to a lower energy orbital. Again, from Fig. 2.7), the emission line transition taking the electron from the $n=4$ to the $n=2$ orbital, requires that the electron loses an amount of energy equal to $E(4)-E(2)$. The electron emits a photon in order to lose the energy required to make the transition. As with the absorption lines, the wavelength of the photon emitted during the emission transition is calculated via Planck's energy formula, with: $E_{4\rightarrow2}=h\,(c/\lambda_{line})=E(4)-E(2)$.

The quantum mechanical rules developed by Bohr and collaborators are an all or nothing scheme. Either the electron absorbs exactly the right amount of energy to make a transition, and then makes the transition, or it stays put. Likewise, an electron either emits exactly the right amount of energy to make a transition, and then it makes the transition, or it stays put. There are no half measures or limbo states where an electron can wait between allowed orbitals in the hope of either gaining or losing a small additional amount of energy to enable a transition.

Furthermore, in addition to the quantization rules for electron transitions, an atom always strives for a distribution of electrons, within the allowed orbitals, that minimizes the total energy. In the case of the hydrogen atom, the simplest and most abundant atom in the entire universe, there is only one electron, and accordingly the minimum energy state condition dictates that, if left to its own devices, the electron within the hydrogen atom will preferentially occupy the $n=1$ ground state. (There is an important catch with respect to this situation that will be discussed in Chap. 4.).

A schematic energy level diagram for the hydrogen atom is shown in Fig. 2.8. By being the simplest (and most abundant) atom in the universe, the hydrogen atom has been well studied by spectroscopists [4]. Those transitions with the electron beginning or ending in the $n=1$ ground state are called Lyman lines, after Theodore Lyman who first studied them in 1906.

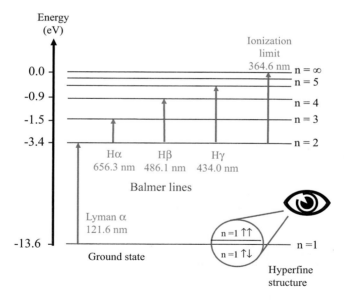

Fig. 2.8 A schematic energy level diagram for the hydrogen atom. The energy axis is expressed in units of the electron volt (eV), where 1 eV is the energy gained by an electron as it moves through an electrical potential difference of 1 V. Accordingly, 1 eV$=1.6022\times 19^{-19}$ J. Lyman lines are those emission or absorption lines associated with the $n=1$ ground state. The Balmer lines are those emission or absorption lines associated with the $n=2$ first excited state [4]. The meaning and astronomical relevance of the hyperfine structure in the $n=1$ state (shown magnified in the *circle*) will be discussed in Chap. 4. Note that the spacing between adjacent energy levels decreases rapidly as the principle quantum number n increases

The first Lyman line (Lyman α), either in emission or absorption, will always be located at a wavelength of 121.6 nm, and accordingly it, and the other lines associated with the $n=1$ state, falls in the ultraviolet region of the electromagnetic spectrum.

Hydrogen lines beginning or ending in the $n=2$ first excited state are called Balmer lines, after Johann Balmer, who first studied them in 1885. The first Balmer line (Hα) transition falls at a wavelength of 656.3 nm, located towards the far red end of the visible part of the electromagnetic spectrum. This line is identified in the 8th panel in Fig. 2.4, and corresponds to Fraunhofer's C line. The Hβ and Hγ transitions have wavelengths that fall in the blue and violet regions of the color spectrum (Fig. 2.4 panel 8, the F and G lines). The Balmer limit corresponding to the $n=2$ to $n=\infty$ ionization transition falls at a wavelength of 364.6 nm, in the ultraviolet part of the electromagnetic spectrum. Historically astronomers have been most interested in the Balmer line series, since many of the transitions fall in the visible part of the spectrum, and they can accordingly be seen with the eye at the eyepiece or photographed in a straightforward manner. Additionally, the spectral classification scheme [2] was developed around the measurable strengths of specific sets of Balmer lines.

Now that the basic quantum mechanical picture of the atom (at least the hydrogen atom) has been outlined, the astronomical relevance of emission and absorption lines can be explained. To produce an emission line spectrum, the electron has first to be placed in a high energy, excited state orbital from which it can later decay to produce some specific emission line, and this can be achieved through collisions with other atoms in a hot gas, or it can be mitigated directly through the absorption of a photon emitted from a nearby star. Likewise, emission lines can be produced under the condition of recombination, where a proton captures an electron into an excited state.

This process, as we shall see below, is particularly important in HII regions. The production of absorption line spectra relies on the fact that an electron in an excited state, in order to move to a lower energy orbital, can emit a photon in any direction. This is especially important in those regions of a stellar atmosphere where spectral lines are formed. In this manner, it is the deep, high-density, high-temperature interior of the star that produces the underlying continuum spectrum, while those atoms in the outer atmosphere pick off just those wavelength photons, traveling outward from the interior, that enable allowed absorption transitions. Non-transition initiating photons simply stream through the atmosphere. Through picking off just those photons that enable transitions, the observer is robbed of their contribution to the continuum, and hence an absorption line is recorded. That is, the energy flux at those wavelengths that enable electron transitions is smaller than would otherwise be the case if there were no atoms in the atmosphere surrounding the continuum-producing interior.

At this stage one might reasonably ask, since electrons in excited states very rapidly drop down to lower energy orbitals, emitting a photon as they go, why is there not an exact balance between the absorbed photons and the emitted ones? The basic answer to this question is that there is a balance, but when the excited electron drops to a lower energy orbital the accompanying photon can be emitted in any direction. It is for this reason that absorption lines are not totally dark, that is, indicating zero energy flux at their specific wavelengths. The isotropic emission property of emitted photons dictates that they are as likely to be directed back towards the interior of the Sun as they are to be directed in an outward direction towards the observer's line of sight. Some of the emitted photons will always end up traveling in the same direction as the continuum photons, but most will not, and it is for this latter reason that absorption lines appear dark, but not so absolutely dark that no energy flux at all is recorded.

With the foregoing technical sections in place we now have enough theoretical detail to move forwards. The properties of light, as a form of electromagnetic radiation, have been described, and the way in which starlight can be interpreted, via spectroscopic analysis, has been outlined. Indeed, it is primarily through the art of spectroscopy and the understanding of the quantum mechanical properties of the atom that astronomers have come to know the stars and learn about the greater universe.

Seeing Red in Orion

The appearance of Galileo's *Sidereus Nuncius* in 1610 is one of those special anchor points in the history of astronomy when something truly new and important happens. The short treatise was a quick, one might say too quick, summary of Galileo's telescopic observations of the heavens. Partly an exercise in self-promotion, in *Sidereus Nuncius* Galileo described new vistas of the celestial realm that no one else had seen before.

Among the unexpected findings that Galileo reported was that of numerous faint stars that were otherwise invisible to the unaided eye. He was also able to resolve into faint stars a number of previously identified nebulous objects. These included the Pleiades (M45) and the nebula in Praesepe (the Beehive Cluster, M44). He also examined the constellation of Orion in great detail. "I had decided to depict the entire constellation of Orion," Galileo writes, but then backtracks on his promise, arguing that there are just too many stars, "for there are more than 500 new stars around the old ones, spread over a space of 1 or 2 degrees." Galileo never actually produced the promised map of the constellation, but he did provide a depiction of the belt and sword of Orion, indicating that he had added 80 new stars to those visible to the naked eye (Fig. 2.9).

Galileo's view of Orion's belt and sword revealed that he had a good eye for detail, but the image and text within *Sidereus Nuncius* leave us with a mystery. Galileo draws no nebulosity. This omission has caused no small degree of debate, with some researchers suggesting that perhaps the nebulosity only became apparent after 1610. Others argue that Galileo simply assumed that the nebulosity would eventually be resolved into stars (consistent with his observations of the Milky Way), and he accordingly omitted the cloud from his diagram. Some give Galileo far too much credit, and it is more likely that he deliberately ignored the nebulosity—probably for at least two reasons. First, he wanted to establish the idea that nebulous stars were really just numerous distant stars seen close together. He probably recognized that a counter example to this—because even through a very small telescope it is clear that the nebula in Orion is more than just a haze of stars—would not help his cause. Second, Galileo knew that there was much ongoing discussion about the possibility that a telescope might introduce *fictitious* features into an image.[5] And again, the last thing that Galileo wanted was a debate about the reality of what he might be seeing. In short, Galileo most likely doctored his image of Orion in an attempt to prove a point and to divert possible criticisms away from his interpretations. In any case, the point appears to be that to Galileo the heavens contained stars and planets only.

---

[5] Certainly early telescope optics suffered greatly from the poor quality of glass available at the time, and it is evident from his drawings of the Moon, as presented in *Sidereus Nunceus*, that Galileo *was* seeing fictitious features.

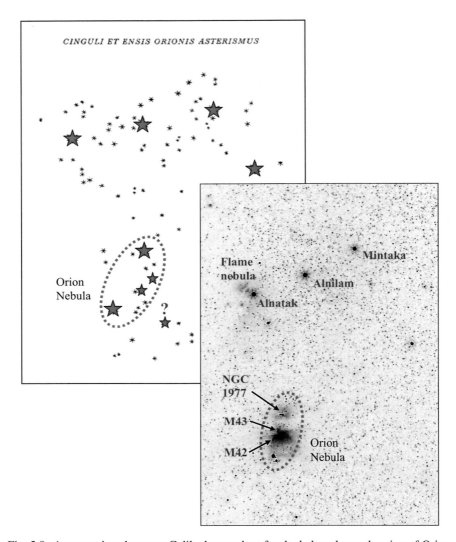

Fig. 2.9 A comparison between Galileo's star chart for the belt and sword region of Orion, as published in his 1610 *Sidereus Nuncius*, and a modern-day photographic image (in negative format). The "known or ancient" naked-eye stars are shown in *red*. The question mark indicates a spurious "known" star. The nebula region in the sword is shown as 4 "known" stars with the addition of some 7 "new" stars in Galileo's drawing

Galileo failed to actually see the nebulosity in Orion's sword, but other observers soon afterwards very clearly saw it. The first known (that is, surviving to the modern era) drawing of the nebula was produced by Giovanni Hodiema sometime before 1654. The Orion Nebula is now one of the great icons of the sky (Fig. 2.10), and no book on astronomy is complete without at least one image of its curtain-like striations and billowing curves.

Fig. 2.10 The iconic Orion Nebula. The distinct *red coloration* in the nebula (and especially Barnard's Loop) is due to the Hα recombination line. The bright star to the *upper left* is Betelgeuse. The *white* nebulosity to the lower right is the Witch Head Nebula (IC 2118). Image courtesy of Rogelio Bernal Andrea. Used with permission

But what exactly is the Orion Nebula? The pioneering spectroscopic analysis by William Huggins, published in 1865, revealed that the Orion Nebula must be a cloud of hot gas.

The presence of the red Hα line in emission, it was eventually realized, betrayed the nebula as being a hot cloud of hydrogen gas. Indeed, the Orion Nebula is an HII (or emission line) nebula. The Roman numeral II in HII indicates that the atom has been singly ionized, and for hydrogen that is the only ionization state possible, since it has just one electron in the neutral HI state. Other elements, with more electrons than hydrogen, can acquire higher ionization states within HII regions, and accordingly emission lines from two, three, and more times ionized atoms are additionally observed.

The strong Hα emission in the spectrum of Orion indicates that we are dealing with a recombination process in which a proton captures a free electron into an excited state. The electron then cascades down through allowed lower energy orbitals towards the ground state. Some of the electron transitions to lower energy orbitals will be directed towards the $n=2$ first excited state, and these will result in Balmer emission lines (such as the distinctive, red, Hα line).

A description of the physical processes occurring in an HII region (as well as those for planetary nebulae—recall Fig. 1.6) was first developed by Danish astrophysicist Bengt Strömgren in 1939. Indeed, building upon ideas presented by Arthur Eddington and the newly available Hα survey observations published by Otto Struve and Christian Elvey in 1938, Strömgren developed a model describing the physical properties of a hydrogen cloud that chanced to envelop a high-temperature star. The resultant Strömgren sphere describes the dynamical balance established between the ionizing effects of the ultraviolet radiation emitted by the star and the recombination that eventually takes place as protons re-capture an electron (Fig. 2.11). High-temperature stars are implicated in the production of Strömgren spheres since, as indicated in Fig. 2.8, to ionize a hydrogen atom (with the electron in the $n=1$ ground state) the energy of the ionizing photon must be at least $E_p=13.6$ eV, and this indicates an associated wavelength of $\lambda_p=91$ nm—i.e., in the ultraviolet part of the electromagnetic spectrum (recall Fig. 2.3).

To produce large quantities of such wavelength photons, Wien's law indicates that an atmospheric temperature of order $T=30,000$ K is required, and this in turn implicates O and B spectral type stars [2], as well as newly formed white dwarf objects, as the ionizing "engines." Accordingly it is found that HII regions are powered by newly formed O and B spectral type stars, while planetary nebula are powered by newly formed white dwarfs.

The size of the Strömgren sphere depends on a whole series of parameters relating to the UV photon flux from the central star, the number density of hydrogen surrounding the star, and the cross-section area of interaction between a UV photon and a neutral hydrogen atom. In short, however, the hotter the star, the greater the UV flux and the larger the Strömgren sphere.

The ionization cross-section area of interaction for neutral hydrogen is $\sigma \approx 10^{-21}$ m^2, and it is this quantity, along with the number density n_H of neutral hydrogen, that determines how far a UV photon is likely to travel before it ionizes an atom. This typical distance will, in fact, be of order $l_{phot} \approx 1/(\sigma\, n_H) \approx 10^{13}$ m (which is just a little more than the diameter of Pluto's about the Sun). Since the photon travels at the speed of light, it will accordingly survive for perhaps several hours before being absorbed in a typical HI

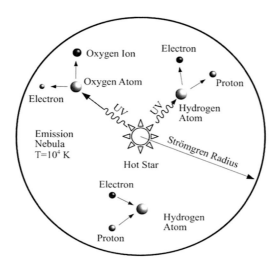

Fig. 2.11 The Strömgren sphere model of an HII region. The boundary of the nebula is determined according to the flux of UV photons from the central star and the surrounding density
of hydrogen atoms. The process at play is that of photoionization followed by recombination.
Accordingly the HII region is first opened up through photoionization with (UV-photon+H
atom) → (proton+free electron). The emission lines are then produced through the recombination process where (proton+electron capture) → (excited H atom → emission lines production
as electron drops to lower energy orbitals). Image courtesy of NASA Cosmos, Tufts University

environment with a hydrogen atom number density of $n_H \sim 10^8$ to 10^9 m^{-3}. Once the core
of an HII region is well established, the interaction cross-section for photons is dramatically reduced and corresponds to the so-called Thomson scattering cross-section
$\sigma_T \sim 10^{-28}$ m^2. Accordingly the photons can move freely through the ionized region and
begin to ionize neutral hydrogen at increasingly greater distances from the central star.

The HII region increases in size not only because of the ionizing effect of the UV photons but also because it is over-pressured. The pressure P of a gas is related to the number
density of gas particles n and the temperature T, such that $P \sim n\, T$. Since the number density of an ionized hydrogen gas (composed of individual protons and electrons) is twice
that of a neutral hydrogen gas (composed of just hydrogen atoms), we have $n_{HII} = 2\, n_{HI}$.
Additionally, the characteristic temperature of the ionized region is $T_{HII} \sim 10^4$ K while that
of the neutral hydrogen is $T_{HI} \sim 10^2$ K, so $P_{HII} \sim 200\, P_{HI}$, and accordingly the HII region will
push outward into the surrounding neutral HI cloud.

Nature only rarely produces a nice Strömgren sphere-like HII region. The factors
controlling the eventual appearance is the total flux of ionizing UV photons, the number
density (along with its spatial variation) in the surrounding HI region, the presence of
more than one ionizing star (as in Orion), stellar winds, and the presence of magnetic
fields. The latter phenomenon appears to play a particularly important role in determining
the complex, but often symmetrical, morphology of planetary nebula.

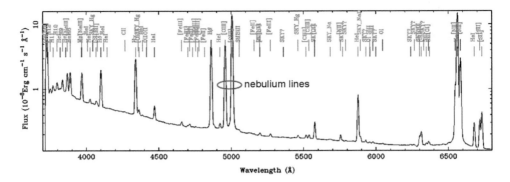

Fig. 2.12 Emission line spectrum for the Orion Nebula. Prominent emission lines are seen for hydrogen (at 6563, 4861, and 4341 Å), twice ionized oxygen (at 4959 and 5007 Å), and singly ionized nitrogen (at 6584 Å). The mysterious nebulium lines were eventually identified with twice ionized oxygen transitions in 1927. Image adapted from S. F. Śanchez et al. www.caha.es/sanchez/orion/

Hydrogen is not the only atom to be ionized in an HII region, and indeed, other atomic elements will undergo the photoionization and recombination process to produce additional sets of emission lines (Fig. 2.12). The nebulium line (discussed in Chap. 1) is one such emission feature prominently seen in the spectrum of the Orion Nebula, and this line is produced by transitions involving ionized oxygen atoms. The Orion Nebula (HII region) is excited by a dominant group of newly formed O and B spectral type stars that make up the Trapezium cluster (so named by Robert Trumpler after the distinct arrangement of the four most obvious, brightest stars—recall Fig. 1.9).

This cluster of some several thousands of stars is believed to be a relatively young Orion feature, having formed as recently as 300,000 years ago. It is the massive stars within the Trapezium that provide the UV photons that drive the photoionization process in the nebula. The most prominent star in the Trapezium, named Theta 1 Orionis C, is actually a binary system composed of stars with masses that are 30 and 10 times greater than that of our Sun. Accordingly, the most massive component (Theta 1 Orionis C1) will undergo supernova disruption in just a few million years.

What a remarkable sight this supernova will be through the telescope—a brilliant jewel sparkling against a sweeping backdrop of gas and billowing dust clouds. Although this vista lies in the deep future, it is apparent that such sights would have technically been available to putative galactic observers in the distant past. The Trapezium is thought to be a sub-component of the larger and much older Orion Nebula Cluster (ONC). Detailed studies of the ONC stars indicate a formation age of some 1–3 million years ago, and the most massive stars within this cluster will have long ago undergone supernova disruption.

At least one or possible several past supernovae explosions in the ONC are the most likely agents underlying the production of Barnard's Loop (recall Fig. 2.10). Named after Edward Barnard, who first photographed, drew, and then described the structure in an 1894 article published in *Popular Astronomy*, the Loop is "an enormous curved nebulosity encircling the belt and the great nebula, and covering a large portion of the giant."

Indeed, almost unable to believe what he was seeing, Barnard writes, "a description of this nebula would not only be complicated but it would fail, also, to give any impression of its form and magnitude." Barnard was correct in his suggestion that an explanation of the curved nebulosity would be complicated, but as the image in Fig. 2.10 happily reveals, a picture can indeed count for a thousand descriptive words.

Barnard's Loop, to say the least, is an enigmatic structure. With a diameter of some 1200 minutes of arc in the sky, as seen from Earth the Loop is estimated to be between 200 and 400 pc away, giving it a physical size of some 60 pc. From the red color due to Hα emission lines, it is clear that Barnard's Loop is an HII region, and various studies indicate that it is kept photoionized by the stars in the so-called Orion OB1 association. Additional recent studies further indicate that the Loop is really the most prominent *visible* part of a vast bubble that has been inflated by multiple supernovae explosions starting some 300,000 years ago. Indeed, the Loop is associated with a region of high velocity high temperature gas called Orion's Cloak.

First described by Lenox Cowie (Princeton University Observatory) and co-workers in 1979, Orion's Cloak is rather odd in that rather than being a cloak of invisibility (in the mode of, say, a dark dust cloud), it is an actual invisible cloak, its presence only discernable through the detailed examination of multiple ionization stages of atoms such as C, N, Si, and S in the line-of-sight direction to background stars. The background stars essentially act like individual searchlights that illuminate small regions of the gas cloud that constitutes Orion's Cloak. Star formation has been taking place within the Orion molecular cloud for perhaps the past 10–15 Ma, and multiple clusters of high mass stars will have been formed and undergone supernova disruption in this time.

Accordingly, Barnard's Loop is a relatively recent structure that is associated with a whole number of nested and expanding supernovae-blown bubbles centered on the Orion Nebula region. The result of all these past supernovae explosions has been the carving out of what is known as the Orion-Eridanus superbubble (Fig. 2.13). This latter cylindrical-shaped region is estimated to be some 350 pc long and about 200 pc wide and was first revealed in survey maps of neutral hydrogen made with radio telescopes in the 1970s (see Chap. 4).

Although the bubble is revealed in neutral hydrogen maps, which essentially highlight the outer bubble boundary, the interior of the bubble is permeated by a high temperature, $T \sim 10^6$ K ionized gas, and thus the interior region emits strongly at X-ray wavelengths.[6] Such supernovae-blown superbubbles are not uncommon within the galactic disk, and they implicate supernovae as the primary movers and shakers of the interstellar medium. Indeed, the Solar System is presently situated within the confines of a supernovae-blown superbubble (identified simply as the Local Bubble).

The Orion Nebula is both a graveyard for past supernovae as well as the birth cloud for newly formed and forming stars. And, while the spectacular death throes of the long-dead massive stars are writ large in space through the production of a superbubble, the birth phase is also evident, and numerous protostars, surrounded by their nurturing accretion disks, have been detected in the Orion Nebula (Fig. 2.14). The details of star formation,

---

[6] This is according to Wien's law, which gives $\lambda_{max} \approx 3 \times 10^{-3}/10^6 = 3 \times 10^{-9}$ m, placing it in the soft-X-ray region of the electromagnetic spectrum (Fig. 2.3).

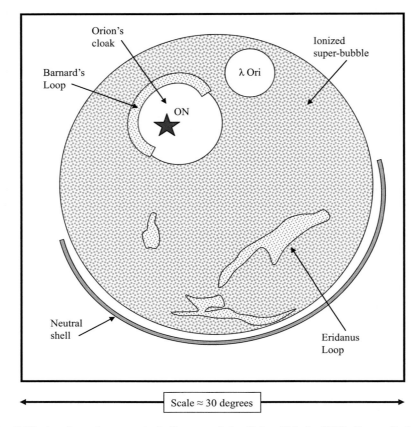

Fig. 2.13 A schematic contextual diagram of the Orion Nebula (ON), Barnard's Loop. Orion's Cloak, and the Orion-Eridanus superbubble. The HII region associated with λ Orionis (Meissa) is also indicated. The image box is approximately 30° on all sides. Image based on a diagram by Bram Ochsendorf of Leiden Observatory et al. Ap.J., 2015

circumstellar disk structure, and planet formation is the topic of Chap. 5, but the Orion protostars are introduced here as examples of structures that can show the consequences of being born in the wrong neighborhood at the wrong time. In this case, the wrong neighborhood is any region close to a massive star.

Again, it is the effect of photoionization that is at play. In this case, however, it is the disks surrounding the newly forming stars that are being gradually eroded by the ultraviolet photons emitted by young, massive, and hot stars. The photo-evaporation process is driven by the heating of the disk gas by surrounding stars, and it limits their lifetime to perhaps a few hundred-thousand years [5]. It is from this relatively short characteristic disk lifetime that the problem for planet formation, in at least some of the Orion protostars, emerges.

Since the stars in the Trapezium have an average age of at least 1 million years, it would appear that any potential planet-forming disks situated close to the cluster will have photo-evaporated long before any planets might have been able to form. Indeed, observations of the Orion Nebula with the Hubble Space Telescope (Fig. 2.14) have revealed the

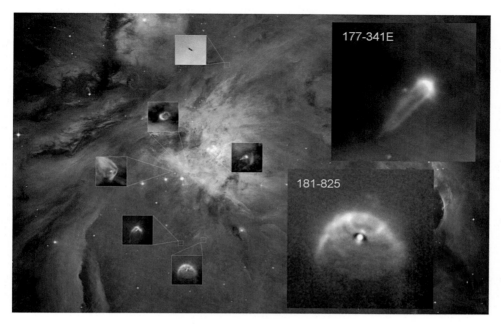

Fig. 2.14 Proto-planetary disks (proplyds) being photoionized by young, massive stars (especially Theta 1 Orionis C) in the Orion Nebula. The bright stars of the Trapezium are revealed in the center part of the image. *Insets to the right* show the proplyds indexed as 181–825 (the Jellyfish) and 177-341E. These two proplyds show distinctive photoionized tails that point directly away from the UV photon supplying star. Image courtesy of NASA/ESA/STSI

photo-evaporation process directly at work. Those systems situated close to the Trapezium show glowing halos of heated gas and long tadpole-like tails pointing away from the cluster. The more remote systems, however, show no such tails or halos, and the disks for these stars are seen as dark silhouettes against the bright nebula background.

It is estimated that the gas and dust disk that produced the planets within the Solar System survived for at least 10 million years, and this suggests that the Sun formed in a smaller birth cluster of stars that than found in the Trapezium. The details will be discussed in later chapters, but the observations indicate that the smaller the total number of stars within a cluster, the smaller the number of massive, high-temperature, and high-luminosity stars they contain. The relative paucity of massive stars further suggests that smaller star clusters are likely to be more efficient at producing low mass stars with accompanying planetary systems.

Tools of the Trade

In terms of the history of astronomy we live in very special times—the epoch in which, for the first time, humanity can measure and comprehend the actual size of the observable universe. Such knowledge requires, beyond the theoretical underpinnings, the development of technologies appropriate to making meaningful measurements. The latter developments refer both to the ability to construct larger and larger telescopes, to see fainter and more distant objects, the design of more and more sensitive detectors that operate across the entire electromagnetic spectrum, as well as the development of new cognitive ways of looking at and thinking about the universe.

Figure 2.15 shows the evolution and development of telescope diameters over time since the early 1600s, while Fig. 2.16 illustrates how the electromagnetic spectrum has been opened up to astronomers over the past century (with a short extension into the near future). Indeed, Fig. 2.16 reveals the incredible advancement in technological abilities and the development, beyond that of the historical optical, of new observational windows by which the universe can be viewed.

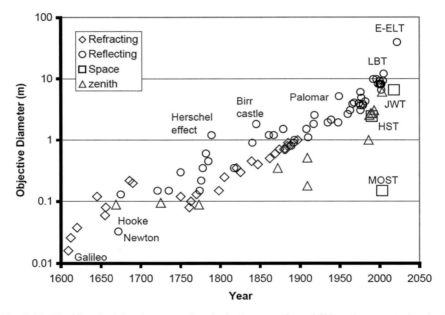

Fig. 2.15 The historical development of optical telescopes from 1609 to the present day (and a little beyond). *Diamonds, circles, triangles, and squares* correspond to refracting, reflecting, zenith, and space-based telescopes, respectively. A number of data points are labeled and indicate Isaac Newton's first refracting telescope of 1668, Robert Hooke's first zenith telescope for 1669, the Birr Castle (Leviathan) telescope of Lord Rosse constructed in 1845, and the Palomar 200-inch telescope of 1949. The 'Herschel effect' indicates the series of increasingly bigger telescopes constructed by William Herschel, culminating in his gargantuan 40-foot reflector constructed between 1785 and 1789. Other telescopes indicated are the Large Binary Telescope (LBT—first light in 2005), the Hubble Space Telescope (HST—first light in 1990), the Microvariability and Oscillation Space Telescope (MOST—operational since 2003). Future telescopes include the James Webb space telescope (JWT—hopefully being launched in 2018) and the European-Extremely Large Telescope (E-ELT—potentially seeing first light in 2024)

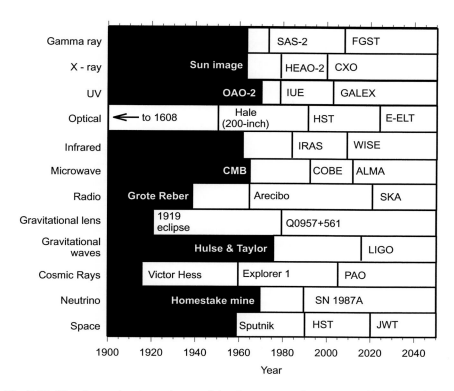

Fig. 2.16 The observational opening up of the electromagnetic spectrum. Also shown in the diagram are the timelines for the development of astronomy through the capture of elementary particles, and most recently that of gravitational wave astronomy. Elementary particle detection astronomy effectively began with the detection of cosmic rays, through balloon borne experiments conducted by Victor Hess from 1911 to 1913. Neutrino astronomy, and the direct probing of the Sun's deep interior, effectively began in the 1970s with the opening of the Homestake mine neutrino observatory

For astronomers, being able to study the universe with eyes that are sensitive across the entire electromagnetic spectrum has been nothing short of revolutionary. The universe that we think we know today is completely unlike the universe familiar to astronomers a mere century ago. In less than three human generations not only has the world view changed, but the cosmic view has changed beyond all recognition and understanding. Indeed, in the October 1923 edition of *Scribner's Magazine*, in an article called "Seeing the Invisible," Nobel Prize winning physicist Robert Millikan noted that virtually all scientific knowledge has been gained at optical wavelengths of light, but then asks, "Have imperceived messages been coming to us in waves outside of this range?".

Although Millikan then goes on to describe how the invisible atomic world had been opened up by experiments and observations exploiting the whole range of the electromagnetic spectrum, he completely failed to see how similar such studies might aid in the study of the greater cosmos. Remarkably, from a modern perspective, Millikan writes that the discovery of radio waves, by Heinrich Hertz in 1886, "has not opened up new worlds to our perception." These words were shown, just 10 years later, to be entirely incorrect—as we shall see in Chap. 4. Indeed, the pace of progress in modern times has been extremely rapid, and the changes often leave the mind reeling with new possibilities.

Although cosmic rays were first identified by Victor Hess, during balloon-borne flights in the early 1900s, it was the development of radio telescopes, initiated in the pioneering work of Grote Reber in the 1940s (see Chap. 4), that jump-started a new era of astronomical research. Furthermore, it was the introduction of space-based platforms, in the late 1950s, that enabled the short wavelength region of the electromagnetic spectrum (γ-rays, X-rays, and the UV) to be explored. Likewise, innovation in detector design during the 1960s enabled the initiation of ground-based infrared and microwave astronomy—the first great triumph of the latter field being the discovery of the cosmic microwave background by Arno Penzias and Robert Wilson in 1965.

In the mid- to late 1970s gravitational lensing techniques were developed[7] following the discovery of the first lensed quasar Q0957 + 561. This technique now allows for the mapping out of galactic dark matter halos as well as the detection of exoplanets. The discovery of the binary pulsar (see later for a discussion on pulsars) PSR 1913 + 16 by Russell Hulse and Joseph Taylor in 1974 expanded the domain of tests on general relativity and can be reasonably taken as the onset of gravitational wave astronomy.

The very first detection of a gravitational wave event was recorded by the Laser Interferometer and Gravitational-Wave Observatory (LIGO) on September 14, 2015. Although the development of the technology required to measure a gravitational wave is incredible in its own right, indeed, it is a great testament to human brilliance,[8] the imagination is left further reeling by the fact that the event itself, which produced a signal that lasted for about 1/100 of a second (shorter than an eye blink), relates to the coalescence of two stellar mass black holes in a faraway galaxy, some 1.3 billion years ago.

The first LIGO detection melds together time, space, spacetime, technology, the history of science, the cosmos, human genius, human imagination, and absolute wonder in one incredible and momentous crossing of world lines. Elementary particle astronomy essentially began in the late 1960s with the opening up of the Homestake mine experiment to study solar neutrinos.

The first extraterrestrial neutrino source, Supernova 1987A, was detected nearly 25 years after the first solar neutrinos were recorded. The recent establishment of the IceCube Neutrino Observatory in Antarctica has taken the field of neutrino astronomy to its next level of development, and, as of 2014, it has been recording about one high-energy astrophysical (that is, non-solar) neutrino per month. Indeed, we are now beginning to see and sense a slow trickle of ghostly particles that promise, eventually, to reveal very deep cosmological secrets [6].

---

[7] The first great experiment relating to gravitational lensing, of course, was that conducted by Frank Dyson (Astronomer Royal), Arthur Eddington, and Charles Davidson during the May 29, 1919, solar eclipse, the lensing effect of the Sun then being used to verify Einstein's newly published theory of general relativity.

[8] There is absolutely no hubris in this statement, and humanity can be deservedly arrogant and boastful about such incredible technological achievements.

Not only do different wavelength detectors reveal different aspects of the universe, they also provide contrasting but telling views of the same vista. The interstellar dust that blocks the view at optical wavelengths can be seen right through by viewing the same scene with an infrared or radio telescope. Likewise, the galactic magnetic field, invisible to optical astronomers, can be mapped out with radio telescopes. The ultra-hot coronal gas that envelopes the galaxy, though invisible to radio and infrared telescopes, is revealed by telescopes and detectors working at ultraviolet and X-ray wavelengths.

Gamma ray telescopes can additionally pick out astronomical features invisible at other wavelengths. Indeed, one such emission line feature of great interest to astronomers is that produced by cosmic aluminum 26, an isotope of aluminum produced during supernovae explosions. It turns out that a gamma ray is emitted during the decay of aluminum 26 to magnesium 26: ^{26}Al → ^{26}Mg + positron + electron neutrino + gamma-ray photon. The emission line is of interest since it can be used to map out the galactic distribution of aluminum 26, and astronomers are interested in aluminum 26 since it was an important radioactive decay heating agent in the early evolution of small Solar System bodies. Indeed, it was the decay of aluminum 26 that provided the energy responsible for the internal sorting of different density materials within the first planetesimals and asteroids.

The fact that ^{26}Al, now seen as ^{26}Mg in meteorites, was involved in the differentiation of asteroids indicates that our Solar System must have incorporated material from a supernova explosion very early on in its formation. Indeed, the gamma ray maps constructed from the ^{26}Al emission line essentially tell us where the most massive stars have formed and where supernovae have occurred within the Milky Way Galaxy during the past several millions of years (Fig. 2.17).

What the astronomer sees, and what underlying physical principles are at play, is shaped and determined by the very eyes (detector) that the astronomer uses. A single picture paints a thousand words, but multiple-spectrum observations of an astronomical object or cosmic vista enable astronomers to write entire volumes.

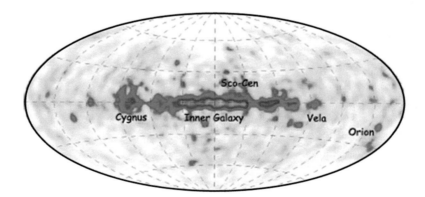

Fig. 2.17   All-sky map of the gamma ray emission line due to ^{26}Al. The clumpy and localized distribution of the ^{26}Al indicates an origin due to supernovae explosions. Several prominent emission line regions are indicated on the map, including the supernova remnants associated with the Cygnus Loop and the Vela Nebula, as well as the massive star formation region of the Orion Nebula. Image courtesy of NASA's Compton Gamma Ray Observatory

Colors of the Crab

The Crab Nebula has a perennial hold on the minds of astronomers, and whenever a new telescope is constructed or a new observational technique is developed one can be sure that M1 will soon be placed on the observing list. Although a brief history of the Crab supernova was presented in Chap. 1, here we look at its many different portraits and multifaceted persona as revealed by telescopes and detectors working across the entire electromagnetic spectrum.

Emission from the Crab Nebula extends over a region amounting to some 7 by 5 minutes of arc on the sky. At a distance of some 6500 light years to the nebula these angular measures indicate a physical region occupying some 11–12 light years in diameter. At the very core of the Crab Nebula, however, is its energy source—a rapidly spinning neutron star that is just 20 km across. We defer the discussion of neutron stars and their appearance as pulsars to a later section. But it is incredible to think that such a small object can power the emission that is seen over a region of space some 10^{38} times larger than its own volume.

Most of the emission observed from the Crab Nebula is due to what is called synchrotron radiation, which is the electromagnetic radiation given off by charged particles (mostly electrons) as they spiral along magnetic field lines (Fig. 2.18). The details of synchrotron emission will be further discussed in Chap. 4. For the present, however, all we need to know is that if synchrotron emission is observed, then we know that there must be some process providing a supply of free electrons and that there must be a magnetic field. Additionally, the power radiated during synchrotron emission is related to the strength of the magnetic field and the energy of the electron. The stronger the magnetic field and the greater the speed of the electron, the greater the amount of energy radiated and the smaller the wavelength of peak emission.

The visual light image of the Crab is iconic, and this portrait is seen almost everywhere (recall Fig. 1.11 and see Fig. 2.19). In the visual the nebula is broadly ellipsoidal in profile, and it reveals a complex trelliswork of intersecting arcs and tendrils. This is the material that formed the outer layers of the original star. The colors in the visual, and especially the outer regions of the nebula, are due to emission lines from ionized hydrogen, helium, carbon, nitrogen, and oxygen. The blue coloration of the inner core region of the nebula is due to synchrotron radiation from relatively low energy electrons. This indicates that there must be an extensive magnetic field structure that permeates the entire nebula. Likewise, the extended emission seen at ultraviolet wavelengths is due to synchrotron radiation from relatively low energy electrons.

Fig. 2.18 The spiral path traced out by an electron as it moves along a magnetic field line. In this illustration the electron has sufficient energy, that is, it is envisioned to be moving at a relativistic speed, that it produces X-ray synchrotron emission. Image courtesy of NASA

←1844 drawing of the Crab Nebula, by Lord Rosse, as seen through his 36-inch telescope. The great, 72-inch, *Leviathan* telescope at Birr Castle is shown to the lower left

Hubble Space Telescope

Astro-1 UV telescope

Fermi γ-ray Telescope

CRAB NEBULA

RADIO | INFRARED | VISIBLE LIGHT | ULTRAVIOLET | X-RAYS | GAMMA RAYS

Very Large Array, NM radio telescope

Spitzer IR telescope

Chandra X-ray telescope

Fig. 2.19 The Crab Nebula as seen with instruments working across the entire electromagnetic spectrum. Six images of the nebula are shown at various wavelengths of light. Key observational instruments and/or satellites are additionally shown, but the portraiture of instruments is far from complete. The Hubble Space Telescope was launched in 1990 and operates across the optical part of the spectrum as well as regions pushing into the ultraviolet and the infrared. The Astro-1 UV telescope was flown aboard the space shuttle (STS-35) in 1990 (the Astro-2 mission flew in 1995), while the Chandra X-ray Observatory was launched during STS-93 in 1999. The Fermi Gamma-ray Space Telescope was launched in 2008, while the Spitzer (infrared) Space Telescope completed it primary mission between 2003 and 2009. Images courtesy of NASA and NRAO

At X-ray wavelengths high energy electron synchrotron radiation dominates, and the view is focused more on the central regions. The X-ray images reveal a condensed central core surrounded by a series of rings and two jet-like prominences. The rings indicate concentrations of emission from material that has been flung outwards by the central neutron star. Time-lapse images of the central core reveal that the strength and even the presence of the ring features change on timescales of order several months. The highest energy gamma ray emission reveals just the central core region surrounding the neutron star. Here we see the very ghost in the machine—the spinning heart of the Crab Nebula.

At gamma ray wavelengths the emission is produced by the very highest energy electrons, traveling at speeds close to the speed of light, in the strongest regions of the neutron star's magnetic field. Remarkably, flares and even superflares, when the emission intensity increases dramatically for several days, are observed at gamma ray wavelengths. These flares are thought to be caused by sudden rearrangements of the magnetic field anchored to the surface of the central neutron star.

Moving to longer wavelengths than optical light, in the infrared part of the spectrum the nebula continues to shine due to low energy synchrotron radiation (seen as red loops), but also shows a contribution due to the thermal emission of warmed dust grains. At radio wavelengths it is synchrotron emission from the lowest energy electrons that is observed, and the nebula takes on a wispy, ghost-like glow.

By observing the Crab Nebula with a range of telescopes and detectors operating at different wavelengths astronomers have been able to examine its entire structure, dissect its interior, and peer into its dark recesses. It is a marvelous yet still mysterious structure. The picture of the Crab has literally been built up piece by piece, and for investigating astronomers some pieces were invisible until they used the correct pair of rose-tinted glasses.

Coronal Gas and a Cosmic Onion

Supernovae are the great thunderbirds of the galaxy. Their wing beats are turbulent, hot-tempered, and full of unconstrained energy. They stir up the interstellar clouds, drive broiling winds, and excavate great cavities that become filled with blistering-hot coronal gas. That the accumulation of supernovae explosions should heat a significant volume of the interstellar medium to temperatures of a million degrees was first predicted before it was actually observed. Indeed, puzzled by how high galactic latitude clouds of neutral hydrogen could survive, Lyman Spitzer (Princeton University) invoked, in a classic research paper[9] published in the *Astrophysical Journal* for July 1956, a very low density, high temperature gas as a constraining agent. At issue, as Spitzer realized, was the problem of gas pressure.

As we saw earlier concerning Strömgren spheres and the development of HII regions, if the pressure inside a gas cloud P_{cloud} is greater than the constraining pressure of the surrounding medium $P_{surround}$, then the gas cloud will push outwards and expand. The existence, as revealed in radio telescope surveys (see Chap. 4), of neutral hydrogen clouds well above the galactic plane was therefore a problem. Why did they not expand and dissipate into space? The answer, as described by Spitzer, was that there is a surrounding medium composed of a very low density, very high-temperature coronal gas.

---

[9] The paper was actually based on a talk delivered at the August 1955 meeting of the International Astronomical Union in Dublin, Ireland.

The word *coronal* is used here in the same manner as it is applied to the Sun's outer atmosphere, or corona, indicating an enveloping, or "crowning," of gas. The coronal gas literally envelops the entire galaxy and essentially fills the otherwise void regions of the interstellar medium, and it establishes a large-scale pressure equilibrium condition. Accordingly, Spitzer argued, the high galactic neutral hydrogen clouds don't disperse because there is a pressure balance between them and the surrounding coronal gas—symbolically: $P_{cloud} \approx P_{surround} = P_{coronal\ gas}$. Again, as introduced earlier, the pressure P of a gas is related to the number density of gas particles n and its temperature T, such that $P \sim n\ T$. Since the temperature T_{HI} of a neutral hydrogen clouds is typically of order 100 K, and the number density n_{HI} is of order 10^8 m^{-3}, the pressure equilibrium condition for interstellar clouds requires that $(n\ T) \approx (n_{HI}\ T_{HI}) \sim 10^{10}$.

Given that the coronal gas must be very hot, with a temperature of order 10^6 K, its density must be extremely small, and the pressure equilibrium condition indicates that the coronal gas must have a density of $n_{corona} \sim 10^4$ m^{-3}. That's about 1 coronal gas atom per a 5-cm sided 'box' of space.

Although Spitzer's idea of a supernova-heated, pressure-constraining coronal gas made theoretical sense, the question persisted as to how it might be directly observed. The answer eventually emerged with the development of space-based UV and X-ray wavelength telescopes. The first indication that the coronal gas was really there was evidenced by the detection of highly ionized absorption lines in the spectra of hot and luminous stars located in nearby Local Group galaxies [7]. Specifically it is the identification of absorption lines due to OVI, NV, and CIV in the spectra of these distant sources that indicates the presence of a very hot intervening gas—the coronal gas enveloping our galaxy.

These highly ionized lines of O, N, and C, which are visible in the ultraviolet part of the spectrum, can only come about in a gas that has a temperature in excess of 10^5 K, and not even the hottest of stellar atmospheres can reach this level of heating. This method of detection acts like a kind of searchlight, with the absorption lines only being seen in the line of sight to particularly bright extragalactic sources. Once X-ray telescopes were placed in Earth orbit, however, from the 1960s onward, the diffuse glow of the coronal gas itself was directly detected (Fig. 2.20).

The stellar searchlight method has been used to map out the structure of the Local (coronal gas-filled) Bubble in which the Solar System presently resides. Some of the most recent work in this area was published in 2010 by a French-American collaboration of astronomers who studied high resolution spectra of some 1857 stars located within a sphere of radius 800 pc centered on the Sun. The team specifically analyzed the characteristics of the ionized calcium K line and a set of lines associated with sodium. The CaII K-line was studied, since it will betray the presence of any partially ionized gas within the line of sight, while the sodium doublet acts as a tracer to any cold, neutral hydrogen in the line of sight.

Here the idea being exploited is that while optical telescopes cannot directly *see* either the coronal gas or the hydrogen gas clouds, the presence or absence of such clouds in the line of sight to specific stars can be inferred through their imprint of absorption line features upon stellar spectra. (This is a direct example of Kirchoff's laws at play; recall Fig. 2.6.) By observing stars around the entire sky, measuring their spectra, and determining their distances from the Sun, a three-dimensional map of the local interstellar medium can be constructed.

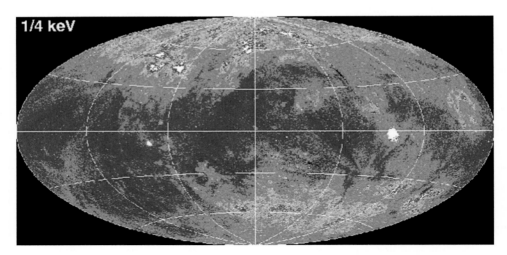

Fig. 2.20 An all-sky X-ray map of the Milky Way based upon ROSAT (Röntgensatellit) data. Although the diffuse X-ray emission is heavily absorbed by dust in the galactic plane (*blue coloration region*), strong but non-uniformly distributed emission (*red, yellow, and white*) is observed in those directions towards the galactic poles. The *white circular region* to the *right* of the image center corresponds to the Vela supernova remnant. The smaller *white region* to the *left* of image center corresponds to emission from the Cygnus superbubble and the Cygnus Loop (Veil Nebula) supernova remnant. Image courtesy of Max Planck Institute for Extraterrestrial Physics

Figure 2.21 shows the deduced distribution of the ionized interstellar medium as traced out by the NaI absorption line features towards the survey stars, and it reveals the Local Bubble by essentially mapping out its enveloping boundary of cold neutral gas. Accordingly, the Local Bubble is estimated to be approximately 160 pc in diameter in the galactic plane. The 3-D geometry of the Local Bubble turns out to be much more complex than that of a simple sphere, however, and in directions above and below the galactic plane it appears to bulge outwards, taking on an hourglass-like figure.

Following the addition of a volume-filling coronal gas component to the interstellar medium it was further realized that there must also be a warm component as well. Writing the seminal reference paper on the topic in 1977, Christopher McKee (University of California, Berkeley) and Jeremiah Ostriker (Princeton) developed what is now known as the 3-phase model for the interstellar medium. Like the Strömgren sphere, the 3-phase model is an idealized abstraction of what is certainly a more complicated reality, but it catches the key ideas of the situation.

In this model the interstellar medium is envisioned as being a series of distinct domains that are all in pressure equilibrium. In the idealized spherical case, like a layered onion, the outermost envelope constitutes the coronal gas with the highest temperature and the lowest number density. In the central core are the neutral hydrogen clouds with the lowest temperature and highest number density. Between the core and the outer envelope resides the warm interstellar medium, which is divided into two main components, depending upon whether the gas is warm and neutral (WNM) or warm and ionized (WIM).

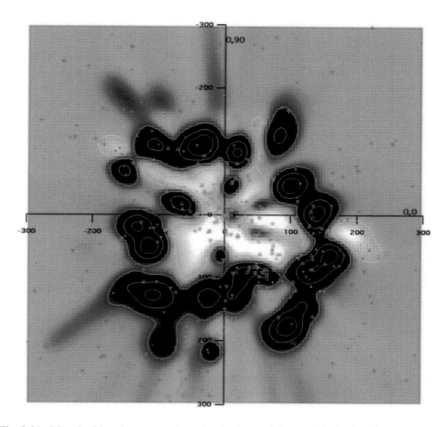

Fig. 2.21 Map, looking down upon the galactic plane, of the partially ionized interstellar gas (as betrayed by stellar CaII K-lines) within 300 pc of the Sun. The *central white region* reveals the very low density Local Bubble, while the *contoured dark regions* reveal the location of the clumpy, highly fragmented bubble boundary. Image courtesy of *Astronomy and Astrophysics*, **510**, A54, 2010

Most of the volume of the interstellar medium is filled by the coronal or hot interstellar medium (HIM), while most of the mass of the interstellar medium resides in the cold, neutral clouds (CNM). The warm neutral and warm ionized regions have temperatures of about 8000 K and number densities of $\sim 3 \times 10^5$ m^{-3}. The enveloping phases of the interstellar medium (the onion structure), therefore, running from the outside in, has HIM → WIM → WNM → CNM. Even though they are not in pressure equilibrium with their surroundings, since they are self-gravitating structures, this nested sequence can be continued to include molecular clouds (MC) and molecular cloud cores (MCC), with CNM → MC → MCC. And, from molecular cloud cores are derived new stars (N✱), with MCC → N✱. The properties of molecular clouds will be discussed in Chap. 4, while star formation is the focus of Chap. 5.

Pulsars and Free Electrons

Supernovae sculpt, control, and determine the dynamics of the interstellar medium, and they also change its composition. This latter process will be explored more fully in Chap. 5, but it is introduced here to remind us that supernovae correspond to the end phase of massive star evolution. The death throes of such stars are rapid, dramatic, and totally glorious, and they are controlled by the extreme physics of shock fronts, elementary particle interactions, quantum mechanics, oppressive density domains, and temperatures well in excess of a billion degrees.

Supernovae are truly beasts born of an incredible alchemy. The compression of the central regions during a supernova event is more than merely brutal, and if not literally pushed out of all existence and transformed into a black hole, the star's once iron core is reconfigured as a neutron star.

The existence of neutron stars was predicted long before they were actually observed. Indeed, the existence of such objects was suggested by Walter Baade and Fritz Zwicky within just a few years of James Chadwick's 1932 discovery of the neutron. As pioneers of modern-day investigations of supernovae, Baade and Zwicky were to write in 1934, "[W]ith all reserve we advance the view that a supernova represents the transition of an ordinary star into a new form of star, the neutron star."

This was a brilliant and brave piece of theoretical reasoning, but it essentially lay forgotten for several decades. The existence of neutron stars was eventually given substance in 1967, however, through the detection of the very first pulsar. Indeed, Antony Hewish, Jocelyn Bell, John Pilkington, Paul Scott, and Robin Collins (all associated with the then Mullard Radio Astronomy Observatory at Cambridge University[10]), in their classic pulsar discovery paper (published in the February 24, 1968, issue of *Nature*), were to write, "a tentative explanation of these unusual sources in terms of the stable oscillations of white dwarf or neutron stars is proposed." In the sense that pulsars are actually pulsating, Hewish and co-workers were soon shown to have outlined an incorrect scenario. The objects are not pulsating but spinning.

The idea that a magnetized and spinning neutron star might be the power source behind the radio emission that was observed from supernovae remnants (and specifically the Crab Nebula) was discussed by Franco Pacini in a *Nature* research paper published in November 1967. Thomas Gold independently invoked a rotating neutron star model in order to describe the pulsed radiation detected by Hewish et al. in a *Nature* research paper published in January 1969.

Neutron stars are not actual stars in the sense of generating energy within their interiors via fusion reactions; rather they are highly compacted objects that are supported against the crush of gravity by a quantum mechanical effect, the so-called Pauli exclusion principal, operating on neutrons. Essentially, neutrons resist the crowding pressure being

---

[10] There has been a sad trend in modern astronomy and physics textbooks to ignore the names of J. D. H Pilkington, P. F Scott, and R. A. Collins when discussing the discovery of the first pulsar. They were an integral part of the research team, and should accordingly be recognized for their role in the discovery [8].

imposed on them through physical compression. Indeed, during the formation of a neutron star the original nuclei in the star's core are so dramatically over-squeezed that their protons and electrons effectively meld together to form neutrons. Thus, neutron stars are essentially massive atomic nuclei.

Under the extreme conditions that apply inside of a neutron star, the exact behavior of matter is not well understood. Nature, however, unbothered by our theoretical ignorance, is able to engineer such objects by compressing up to as much as 3 times the mass of the Sun into a volume of space no larger than 20 km across.

In addition to their extreme density and small size, neutron stars can also support and maintain extremely strong magnetic fields, and it is the rotation of this magnetic field, controlled by the spin of the neutron star, that drives the pulsing effect (Fig. 2.22). The standard pulsar model is that of the oblique rotator in which the spin axis of the neutron star and the axis of its associated magnetic field are not aligned. Accordingly a form of lighthouse effect comes into play with the magnetic poles, where the electromagnetic radiation is actually generated, being swept around the sky.

In this manner the magnetic poles are briefly, but periodically, brought into the direction of the observer's line of sight. Essentially the emission regions are only observable when the magnetic poles are directed towards the observer, and it is the spin of the neutron star that modulates how often the pulses will be recorded. The first pulsar discovered by Hewish et al. has a modulation period of 1.337 s, with each pulse lasting 0.3 s[11]; the neutron star at the heart of the Crab Nebula (recall Figs. 1.11 and 2.19) pulsar spins an incredible 30 times per second.

By analyzing pulsar signals one can begin to study both the physical structure of neutron stars as well as the extreme conditions under which the emission process works. It also turns out that a detailed analysis of the signal arriving from a particular pulsar is highly useful for the information it carries about the ionization state of the interstellar medium. Specifically a feature known as the dispersion measure $DM = n_e D$, where D is the distance to the pulsar, can be quantified, and this measure informs us about the number density of free electrons n_e in the direction of the pulsar. The dispersion measure is associated with an observed delay Δt in the pulsar's signal arrival time, and this delay correlates with frequency (Fig. 2.23). While Δt can be measured from the dispersion data (Fig. 2.23), the frequency is set by the radio telescope receiver. So, if a distance to the pulsar can be determined,[12] a value of n_e can be derived. Additional data in the form of the emission measure EM, which varies according to the electron density squared, and the Faraday rotation measure RM, which varies with the electron density and the galactic magnetic field along the line of sight, is also available from the analysis of pulsar data.

---

[11] It is staggering to think about this incredible object (and its kin). PSRB1919+21 has a mass about 1.4 times larger than that of our Sun, a radius that is over 71,000 times smaller (r ~ 10 km) than our Sun, and it spins on its axis 4 times every 3 s. That such objects exist in nature is truly incredible!

[12] This in fact is no easy task. A few pulsars have known parallax distances (recall Fig. 1.14), but most have to be estimated by secondary association. Since neutron stars are products of supernova disruption, so the distance to a pulsar can be estimated by determining the distance to its associated supernova remnant (see Fig. 1.12).

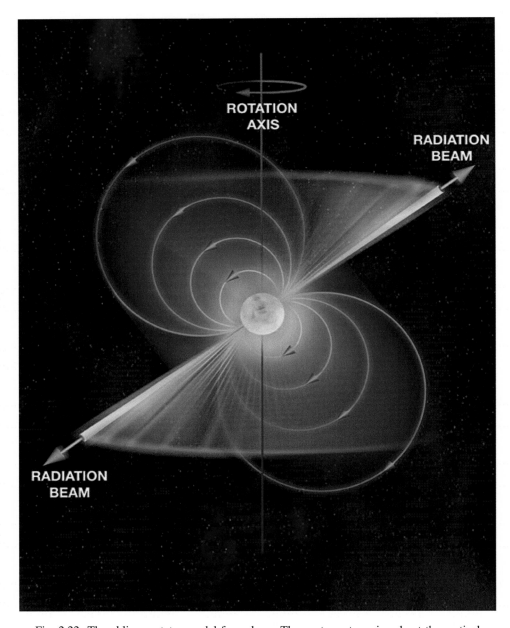

Fig. 2.22 The oblique rotator model for pulsars. The neutron star spins about the vertical axis, while the radiation beams are aligned along the magnetic field axis. A pulse of radiation is recorded when the magnetic poles are alternately spun into the observer's line of sight. Image courtesy of Bill Saxton/NRAO/AUI/NSF

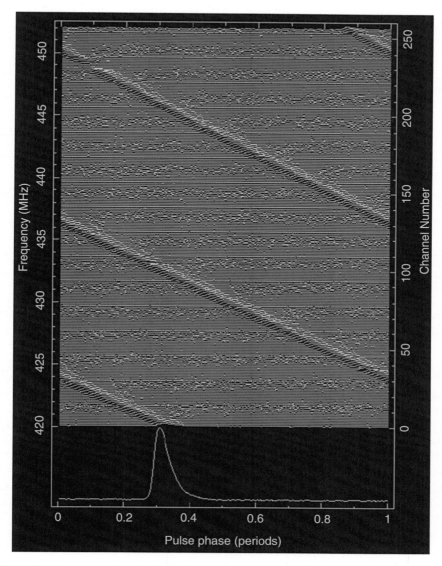

Fig. 2.23 The systematic variation in the arrival time of a pulsar signal is revealed through the diagonal ridges in this frequency versus time diagram. The de-dispersed profile, when the time delays are all removed, is shown at the *bottom* of the image. Image courtesy of Jodrell Bank Observatory, University of Manchester

By bringing all the *DM, EM,* and *RM* data together a line of sight map of the free electron number density of the galactic disk can be constructed. Typically it is found that $n_e \sim 20{,}000$ per cubic meter, although the number can vary significantly across the galactic disk.

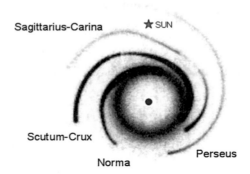

Fig. 2.24   The spiral arm structure of our galaxy as constrained by pulsar dispersion measure mapping of the free electron number density. Near the Sun (*red star*) $n_e \sim 2 \times 10^4$ m^{-3}, while in the outer regions of the Scutum-Crux spiral arm, $n_e \sim 2 \times 10^5$ m^{-3}. The Sun is located in the so-called local spur associated with the Sagittarius-Carina spiral arm. The *central circle* corresponds to the 5 kpc ring (see Chap. 4), while the *blue dot* indicates the location of the galactic center. Image adapted from Taylor and Cordes, ApJ. **411**, 1993

This pulsar based dataset can be further combined with the observations relating to the spatial distribution of HII regions, young galactic clusters, young massive stars, and HI clouds (see Chap. 4) to map out the distribution of matter within the galaxy. Indeed, such survey data reveals the spiral arm nature of our Milky Way Galaxy (Fig. 2.24).

The model illustrated in Fig. 2.24 is taken from the now classic paper by Joseph Hooton Taylor[13] (Princeton University) and James Cordes (Cornell University) published in 1993, and though it is based upon the pulsar dispersion measure data it effectively traces out the WIM distribution of the interstellar medium. The model for the spiral arm structure in our galaxy, as revealed through the electron number density data, is markedly nonsymmetrical. A distinct flattening, for example, is evident in the Sagittarius-Carina and Scutum-Crux spiral arms in the region between the Sun and the galactic center.

There has been some debate in recent years as to the actual number of spiral arms that exist within the disk of the Milky Way. Data gathered under the guise of the Two Micron All-Sky Survey (2MASS), conducted in the late 1990s, for example, appeared to indicated that there were only two spiral arm features rather than four. This result, however, is now thought to be an artifact of the survey data. Indeed, the 2MASS survey, observing in the infrared part of the spectrum, was mostly picking out the low mass, low luminosity stars within the disk of the galaxy, and since these stars are long-lived (see Chap. 5) they have

---

[13] This is the same Taylor in the Hulse and Taylor binary pulsar that enabled the first verification of Einstein's prediction concerning gravitational waves. Russell Hulse and Taylor were awarded the 1993 Nobel Prize for Physics for their discovery of PSR1913 + 16 (recall Fig. 2.16).

moved well away from their star-forming clouds. Accordingly, these older low mass stars have dispersed into the inter-arm regions of the disk, blurring the distinction between inner and outer arms delineated by the massive, short-lived but highly luminous stars, the HII regions, young clusters, and pulsars.

REFERENCES

1. Interested readers will find more details on Maxwell's ideas and the properties of electromagnetic radiation in the author's book, *The Physics of Invisibility* (Springer, New York, 2012).

2. The spectral classification scheme was developed over several decades encompassing the turn of the 19th century. Literally hundreds of thousands of stellar spectra were recorded photographically, and numerous teams of technicians and astronomers sorted through the data looking to tease out an order sequence. Spectra were ordered in terms of line strengths (literally the area of a spectral line) and line strength ratios, and these in turn were interpreted in terms of a temperature sequence. Stars are now classified according three parameters: a spectral type, a sub-group number, and a luminosity class. There are seven principle spectral types, represented by the letters O, B, A, F, G, K and M, with the O stars being the hottest and the M stars the coolest. The sub-group number ranges from 0 to 9, with 0 representing the lowest temperature within the type and 9 the highest temperature. The luminosity class is given as a Roman numeral between I and V. Luminosity class I applies to the supergiants, which are the largest and most luminous stars; luminosity classes II, III and IV are the intermediate giant stars, while luminosity class V corresponds to the lowest luminosity dwarf stars. By this scheme, the Sun is a G2 V star. See table A1 in Appendix I for further details.

3. Interested readers will find more details on the experiments and development of theoretical ideas concerning atomic structure in the author's book, *The Large Hadron Collider – Unraveling the Mysteries of* Universe (Springer, New York, 2010).

4. The energy associated with each orbital in the Bohr model of the hydrogen atom has a particularly straightforward expression, being $E(n) = - R_H / n^2$, where R_H is the Rydberg constant of 13.605 eV (see caption to figure 2.8 for a definition of the electron volt). Accordingly the energy difference between two electron orbitals having principle quantum numbers n and m (with $m > n$) is $\Delta E = E(m) - E(n) = R_H (1/n^2 - 1/m^2)$. It is the amount of energy ΔE that an electron has to either absorb (to enable a transition from n to m), or emit (to enable a transition from m to n).

5. Specifically, the UV photons set up a high temperature, $T \sim 10^4$ K, ionization front that heats the outer layers of the disk, and as a consequence of this the sound speed $c_S \sim (T/\rho)^{1/2}$, where ρ is the density, associated with the disk gas is increased to about 10 km/s. For a central protostar of mass M, the radius R_{esc} beyond which the sound speed exceeds the system escape velocity V_{esc} is $R_{esc} = 2 G M / (c_S)^2$, where G is the universal gravitational constant. The radius at which $V_{esc} = c_S = 10$ km/s for a 1 solar mass star is about 18 astronomical units. Accordingly, for radii $R > R_{esc}$ the gas is able to escape from the disk in the form of a thermal wind. Observations indicate that in the Orion Nebula the thermal wind can drive a disk mass loss rate of $M_{wind} = 10^{-7}$ $M_\odot$/yr, and given a typical disk mass $M_{disk} = 0.01$ $M_\odot$, the disk lifetime is limited to $t_{disk} \sim M_{disk}/M_{wind} \sim 10^5$ years.

6. The current grail-quest in the field of particle physics is that of finding the carrier of dark matter. That such matter, which neither absorbs nor interacts with electromagnetic radiation but has a very definite gravitational influence, exists is beyond question. What is not known yet is what kind of particle (or particles) is involved. Additionally, it is not known where the dark matter particles fit within the present plethora of possibilities offered by the beyond-the-standard models of particle physics. Interested readers will find more details on the past and present experiments relating to dark matter detection in the author's book, *The Large Hadron Collider – Unraveling the Mysteries of* Universe (Springer, New York, 2010).

7. The Local Group of galaxies is composed of about 50 galaxies spread over a region several millions of parsecs in extent. The Milky Way and the Andromeda galaxy are the most massive members of the group and they, along with their many satellite galaxies, are located towards the system center. The Local Group is part of the much more extensive Virgo Supercluster.

8. There has additionally been a ridiculous amount of bluster written about the supposed denial of a Nobel Prize to Jocelyn Bell because she was a *mere* graduate student and/or a woman. Neither charge is remotely true. The Nobel Prize for Physics was awarded to Antony Hewish and Martin Ryle in 1974 for, "their pioneering research in radioastrophysics". The award was not made for the specific discovery of pulsars.

3

The Dark Clouds Revealed

Kapteyn Tries Again

Jacobus Cornelius Kapteyn was the quintessential lone-wolf astronomer—brilliant, determined, highly capable, but not one of the established elite. Born in 1851, the tenth of fifteen children born to his parents Gerrit and Elisabeth, Kapteyn was thrust into a world of dynamical chaos that only a large family can provide,[1] and perhaps therefore it is no surprise that he ultimately became an expert on the application of statistical methods to astronomy.

Having completed his doctorate related to the vibrations displayed by thin membranes at the University of Utrecht in 1875, Kapteyn changed scientific directions and joined the observational staff at Leiden Observatory. In 1878, however, at the tender age of 27 years, Kapteyn took a position at the University of Groningen as Professor of Astronomy and Theoretical Mechanics. It was at Groningen that Kapteyn encountered a somewhat awkward problem for a new professor of astronomy—no telescope or observatory. Indeed, his initial laboratory space was to be a room borrowed from the physiology department. Undaunted, Kapteyn launched into a program of measuring photographic plates to determine stellar proper motions and parallax distances.

In 1877, while at Leiden Observatory, Kapteyn published his first astronomical paper, a joint effort with Hendricus van de Sande Bakhuyzen, which concerned position and brightness reductions for comet Winnecke. It was the appearance of another comet, the Great Comet of 1882, however, that ultimately gave direction to Kapteyn's life's work. The remarkable Sun-grazing comet of 1882 was monitored by the great southern observer (Sir) David Gill, then employed as Her Majesty's Astronomer at the Cape of Good Hope in Africa, and who, almost on a whim, decided to try to photograph the comet.

---

[1] Kaptyen's parents also ran a boarding school for boys, where Kapteyn was well educated but later confessed that the vocational demands on his parents left him feeling alone, isolated, and neglected.

© Springer International Publishing AG 2017

M. Beech, *The Pillars of Creation*, Springer Praxis Books, DOI 10.1007/978-3-319-48775-5_3

Borrowing a portrait camera from a local photographer, Gill set the device on the telescope tube and subsequently captured a number of remarkable images of the comet's tail. More importantly, however, Gill captured on these same photographic plates the images of thousands of stars, and it was this observation that suggested to him the possibility of producing a photographic atlas of the southern sky.

Although such an atlas would be an invaluable complement to Friedrich Argelander's northern hemisphere *Bonner Durchmusterung*, Gill soon found that the effort needed to analyze the photographic plates was beyond the resources of the Cape Observatory. It was upon hearing of Gill's difficulties that Kapteyn made the move that would map out the course of his astronomical career. He contacted Gill and offered to perform the analysis measurements of the photographic plates. The reduction of a photographic plate is a time-consuming process, and although the positional measurements are made with a plate-measuring engine, great quantities of tedious calculation are required to obtain a final result. Working with a minimal staff, including inmates from the local jail, Kapteyn spent 10 years, from 1886 to 1896, reducing the positional data on the 454,875 stars brighter than the 11th magnitude that constitute the three-volume set. Published between 1895 and 1900, the *Cape Photographic Durchmusterung* was a monumental achievement, and it successfully launched Kapteyn as a major player onto the international astronomical scene.

In addition to analyzing the Cape Observatory plates, Kapteyn worked on stellar parallax measurements, and he began to develop statistical techniques that would enable the study of large-scale stellar motions. Indeed, it was through the study of proper motion data relating to many thousands of stars that Kapteyn made a remarkable and unexpected discovery. Publishing his initial findings in 1904, Kapteyn found that rather than stellar proper motions being randomly distributed in the sky, as expected up to that point, they were, in fact, preferentially directed towards and away from the galactic center along two streams running approximately 140° apart from each other.

Kapteyn initially thought that this discovery implied the existence of two star streams moving in opposite directions around the galactic center. However, this was too literal an interpretation, and it was later realized by Jan Oort (one of Kapteyn's students, in fact) that the proper motion data indicated not that there were two distinct star streams but rather that the galaxy as a whole was rotating. Indeed, what the proper motion statistics really showed was that all stars are traveling in the same direction around the galactic center but with speeds that vary according to galactocentric distance.

Building on the ever growing interest in star statistics, Kapteyn realized that with the help of observatories from around the world it should be possible to explore the nature of the galaxy through star counts—just as William Herschel had tried to do in the late eighteenth century. In 1906 Kapteyn set his plan into motion and mapped out 252 1-square degree regions on the sky. He then enlisted the help of as many observatory astronomers as he could reach, asking them to study the stars in the selected regions, measuring their magnitudes, spectral types, proper motions, and radial velocities. It was to be a monumental effort, conducted by many astronomers, and it was the first truly global astronomical observation program.

Over the ensuing 15 years data began to amass, and it was in 1922 (a year after his official retirement from Groningen) that Kapteyn published in the *Astrophysical Journal* a lengthy summary paper entitled, "First attempt at a theory of the arrangement and motion of the sidereal system." In this highly mathematical paper Kapteyn brought together truly vast amounts of data on star counts, and he began to set out his thoughts. He deduced that the observed sidereal system contained some 47 billion stars, all of which were in rotation around the galactic polar axis.[2] The Kapteyn universe was essentially lens-shaped, being some 17,000 pc across and 3500 pc at its thickest. The Sun was apparently not located at the center of the star system but was displaced away from the central spot by some 650 pc.

Kapteyn brought together a truly remarkable amount of data, but his results turned out to be entirely spurious. The fatal error in his analysis was that of ignoring the absorption of starlight by interstellar dust, indeed, the very same issue that had defeated William Herschel when interpreting his star-gauging data. Kapteyn was certainly aware of the effect that interstellar dust might have on his analysis. It was just that he greatly underestimated how much dust there might be in the galactic plane.

Although Kapteyn's universe is now considered an interesting historical sidelight, and an early but problem-plagued attempt at the statistical analysis of stellar positional data, Kapteyn's work was certainly not in vain. Kapteyn was a pioneer in the field of stellar statistics. He was a pioneer in the art of initiating global cooperation in astronomical data gathering, and he was mentor to many of the key players, the Dutch school of astronomers, as it were, who would eventually produce the present-day detailed picture of galactic structure.

A System of Stars, or Many Star Systems

That some of the nebulous clouds could display a distinct spiral structure was first realized by William Parsons, Third Earl of Rosse, in 1845. Using his 72-inch. Leviathan telescope of Parsonstown, he swept up the Whirlpool Galaxy (Fig. 3.1), and he sketched its distinct, grand-design spiral arms and interacting companion. Indeed, it is now known that the Whirlpool Galaxy is really two gravitationally interacting galaxies—the smaller component being a highly distorted dwarf galaxy.

Located some 7.1 million parsecs (23 million light years) away, M51 is not the closest spiral galaxy to the Milky Way. Our closest such companion is another long-observed nebula, the Andromeda Galaxy (Fig. 3.2). Located some 778 pc (2.54 million light years) away, the Andromeda Galaxy is the most distant object visible to the unaided human eye. Charles Messier first recorded the Andromeda "nebula," designated M31 in his catalog, in 1764, although its strangely elongated appearance in the sky had been noted long before this time. Though we currently recognize M31 and M51 as distinct objects, galaxies in their own right, it took astronomers a good while to agree that this was indeed the case.

The pivotal question at the turn of the twentieth century was whether the spiral nebula were really remote systems of stars, independent of the stars that constitute the Milky Way, or were they parochial clouds of gas and dust situated within the Milky Way, with the Milky Way being the overarching domain of all stars? The only way to settle the debate was to determine distances.

---

[2] Kapteyn still believed in his two counter rotating star streams at this stage.

Fig. 3.1 The grand design spiral arm features of the Whirlpool Galaxy (M51) are beautifully portrayed in this Hubble Space Telescope vista. The interacting dwarf companion NGC 5195 is seen to the *right* in the image. Charles Messier first cataloged the faint glow of the Whirlpool Nebula in October of 1773, and indeed, it is number 51 in his famous catalog. Image courtesy of NASA/HST

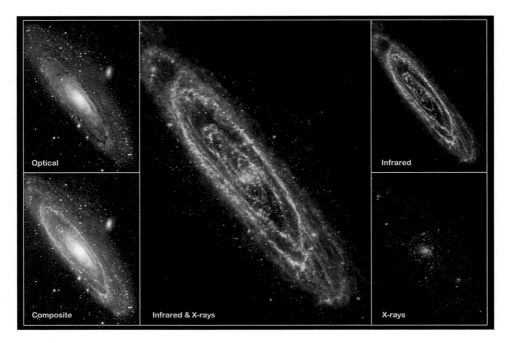

Fig. 3.2 The Andromeda Galaxy (M31) imaged at optical, infrared, and X-ray wavelengths. The spiral arm dust features within the disk are clearly seen in these images, along with the high-energy, X-ray sources that cluster towards the galaxy center. Image courtesy of ESA/Herschel/XMM

The first salvo in the debate concerning spiral nebula distances was fired by Heber Curtis. Importantly, Curtis, working at Lick Observatory, identified a nova that was clearly situated in M31, and because such objects had been calibrated for brightness (more specifically luminosity), he was able to estimate that the Andromeda nebula was at least 500,000 light years away. Although an underestimate by modern standards, the derived distance to M31 convinced Curtis that the nebula must be an independent system of stars, indeed, a galaxy of stars that was distinct from the Milky Way.

The cosmos for Curtis was accordingly filled with massive individual galaxies, our own Milky Way being just one of many. In support of his argument, Curtis also drew attention to the dark bands seen across edge-on spiral galaxies (Fig. 3.3). These bands, he argued, were analogous to the dark clouds resident in the disk of our galaxy. Meanwhile, however, as Curtis was presenting his results, Harlow Shapely, at Mount Wilson Observatory, had other ideas about the structure of the cosmos.

Shapley's observational approach was different from that adopted by Curtis. Specifically, Shapley studied globular clusters, and, using a period-luminosity relationship that had been calibrated for RR Lyrae stars [1], he reported in 1918 that the most distant globular cluster (NGC 7006) was at least 220,000 light years distant. Furthermore, Shapley took note of the non-symmetric distribution of globular clusters in the sky (recall Chap. 1 and Fig. 1.19a). Indeed, he argued that the almost swarm-like globular clusters must be in orbit

Fig. 3.3 The edge-on spiral galaxy NGC 4013, located some 61 million light years (19 million parsecs) away, clearly showing the strong absorption of starlight by interstellar dust situated within the plane of the galactic disk. Image courtesy of Wisconsin-Indiana-Yale-NOAO [WIYN] telescope, Kitt Peak

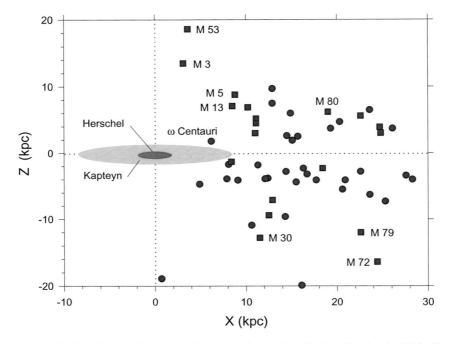

Fig. 3.4 The distribution of globular clusters as deduced by Harlow Shapley in 1918. The asymmetry in the sky location of globular clusters is clearly seen, and Shapley used this observation to deduce that the Sun was located some 13 kpc (42,000 light years) from the galactic center (GC). Note that by modern standards, Shapley overestimated the distance to the globular clusters and the galactic center. Indeed, the present canonical value for the distance of the Sun from the GC is 8 kpc. The approximate dimensions of the stellar universe as deduced by William Herschel (1811) and by Kapteyn (1922) are additionally shown in the figure—centered on the Sun at the (0, 0) location. Data for the globular cluster distribution from the *Astrophysical Journal*, 48, 154, 1918

around the central regions of the Milky Way (in the direction of Sagittarius), and he used the deduced distance to the clusters to argue that the Sun must be displaced by about 42,000 light years (~13 kpc) from the center of the galaxy (Fig. 3.4). The picture promoted by Shapley, in effect, was that the entire universe was essentially full of stars, star clusters, and small nebulae, there being no such objects as remote galaxies or island universes.

So, is the universe one big galaxy, that is, a single large system of scattered stars, or is it full of individual galaxies (island universes composed of billions of stars) that are widely separated from each other? The debate, so to say, on which interpretation was the right one—that by Shapley or that by Curtis—came to a head at a meeting held at the Smithsonian Museum of Natural History in Washington on April 26, 1920.

There was no contentious debate between the combatants. Both simply stated their viewpoints and summarized their specific observations. The details presented by Shapley and Curtis during the Smithsonian meeting were summarized, with each adding rebuttals to their opponent's viewpoint, in two papers that were published in May of 1921. After that it was basically up to the reader to decide who had presented the most coherent and believable picture. It is probably safe to say that no one person specifically won the verbal debate (so called); rather, it was further observational discoveries, by other astronomers, that settled the argument. The two key players were Vesto Slipher (Lowell Observatory) and Edwin Hubble.

When it came to recording and measuring the spectra of faint nebula, Vesto Slipher was the great innovator and practitioner. He toiled for many hours at the telescope in order to acquire useful results, and with respect to the Andromeda nebula (M31) his efforts paid off in 1913. Writing of his results in the *Lowell Observatory Bulletin* (# 58) Slipher announced the discovery of a Doppler shift in the absorption lines of M31 that amounted to a radial velocity of 300 km/s. This was an astounding result, since it revealed the nebula had an associated speed ten times greater than that found for any then known star.

In 1921 Slipher published Doppler shift velocities for the galaxies NGC 584 and 936, finding staggering (for the time) recession velocities of 1800 and 1300 km/s, respectively. The implications were clear; if all spiral nebulae were moving with the characteristic speeds being found by Slipher, then they could not possibly be local structures, else a true proper motion shift would soon be realized. Such fast-moving objects would only appear as fixed features in the sky if they were additionally located at vast distances away in deep space.

Slipher had shown that spiral nebulae had incredibly high radial velocities, but it was Edwin Hubble, working at Mount Wilson Observatory in California, that provided the evidence to indicate they were also located at incredible distances. In many ways, Hubble's immensely important result simply leaked out to the outside world. Hubble had let other astronomers know of his results by word of mouth and in letters, and *The New York Times* newspaper even reported his conclusions in their November 23, 1924, issue, over a month before the first official announcement was made.

Even then, what might be called the official announcement hadn't been specifically planned but was a last-minute contribution to the American Astronomical Society meeting being held in Washington, D.C., from December 30, 1924, to January 1,1925. Hubble was not actually at the meeting, and his paper was read out to the audience by Henry Norris Russell. The paper was concerned with the analysis of light curve data for Cepheid variables that had been located within the Andromeda and Pinwheel galaxies (M31 and M33, respectively), with the specific result that each galaxy was found to be of order 930,000 light years away. At this distance the Andromeda Galaxy, Hubble calculated, must be at least 30,000 light years in diameter—larger even than the Milky Way as deduced by Shapley.

The implications of Hubble's study were clear. The spiral nebulae were located at such vast distances that they could only be island universes, or as we now call them, spiral galaxies [2]. The one big galaxy model of Shapley was accordingly wrong, and the individual galaxy model of Curtis was right. In the end, to some extent, however, both combatants in the Great Debate were right. Curtis gave us the modern viewpoint that the universe is full of galaxies, whereas Shapley developed the means (once refined) to determine the greater structure of the Milky Way Galaxy.

Trumpler's Dim View

The problem with dust is that it is usually difficult to see, and yet, just as is any room at one's home, a close examination reveals that dust is everywhere—not just in the obvious places, such as on the tops of books and treasured ornaments but in all those hidden crevices we normally ignore, such as the back of the television and behind the refrigerator. So to with our galaxy. Interstellar dust is everywhere.

In many regions along the Milky Way the dust is concentrated to such a high level that it creates what appear to be dark voids (recall the Coalsack Nebula in Fig. 1.5). Such regions are not real voids, or regions totally empty of any stars, but they are regions where the absorption of starlight by interstellar dust is extreme. This situation was first hinted at by Edward Barnard in the early twentieth century. Working from Yerkes Observatory in Chicago, Barnard obtained photographic images of many dark nebulae, and in late 1913 wrote in the *Astrophysical Journal* that, "the so-called 'black-holes'[3] in the Milky Way are of very great interest. Some of them are so definite that, possibly, they suggest not vacancies but rather some kind of obscuring body lying in the Milky Way, or between us and it, which cuts out the light from the stars."

A few years later, in 1919, Barnard published, again in the *Astrophysical Journal*, a catalog of 182 "dark markings of the sky" recorded photographically. Although Barnard argued that most, but not necessarily all, of these dark markings must be "obscuring masses" in this publication, he declined to speculate on the nature of the obscuring matter.

William Herschel, in the early nineteenth century, knew nothing of interstellar dust, and his star count data accordingly indicated a relatively small, disk-like distribution for the stellar realm (Fig. 1.7). Indeed, the bifurcation in his stellar disk is entirely due to the absorption of starlight by interstellar dust in the direction of the center of our Milky Way Galaxy.

Jacobus Kapteyen, in the 1920s, was fully aware of the dimming effects of interstellar dust, building as he was on the speculative results of Barnard, but he chose to ignore the issue when analyzing his star counts. Accordingly Kapteyn, just like Herschel before him, discerned that the stellar realm had a disk-like structure with the Sun located close to the central point (Fig. 3.4). In each survey the limit to which stars could be seen was entirely set by the absorption and scattering of starlight by an all-pervading smog of interstellar dust in the plane of the Milky Way. Certainly, interstellar dust can clump into very obvious dark assemblages (Barnard's "Black-Holes"), but even when the view is apparently pristine and clear there is still a vast sea of interstellar dust along the observer's line of sight, and little by little, the greater the depth along the line of sight, the greater the accumulated effect of the dust.

How, then, to demonstrate that interstellar dust really does pervade the entire disk of our Milky Way Galaxy? One way to achieve this, and indeed the method adopted by Robert Trumpler in the late 1920s, is to compare distance estimates to (reasonably) well-calibrated objects by two distinct methods—one method revealing a distance estimate based on the observed brightness, d(light), and the other giving a distance based upon pure geometry, d(ruler). The point here is that the first method will be affected by the dust in the observer's line of sight, while the latter will not. If there is no dust and/or absorption of

---

[3] These are not the gravitationally collapsed black holes of modern general relativistic interest. Barnard is simply applying a visually descriptive term.

starlight in the line of sight then d(light) must equal d(ruler), but if there is absorption of starlight then invariably, d(ruler) will be smaller than d(light). Essentially the absorption and scattering of starlight by interstellar dust makes stars appear both redder in color (a process called reddening; see below) and further away.

In his classic analysis Trumpler chose to look at galactic clusters (Fig. 3.5). He reasoned that such clusters could be calibrated according to the brightness of their most luminous stars and according to their angular size. As we saw earlier, the distance to an object can be obtained from the measure of its angular size α in the sky and by assuming a calibrated physical diameter. This is the standard ruler method described in the caption to Fig. 1.12. Under this methodology the angular size of a galactic cluster acts as a proxy for its distance in that the further a cluster is away from us, the smaller will its angular size appear.

With the d(ruler) distance measures determined, Trumpler then set about finding the distances to the same sample of clusters according to characteristics of their brightest stars. This method, which provides the d(light) measure, relies on measuring the spectral type (see the Appendix in this book) and the apparent visual magnitude of the brightest stars in each cluster. The measured apparent magnitude can then be compared with the (previously) calibrated absolute magnitude appropriate to the spectral type of the star being observed, to reveal d(light).

Fig. 3.5 The Pleiades Cluster (M45). Visible to the unaided human eye as a small smudge of luminescence on the night sky in the constellation of Taurus, the Pleiades is a galactic cluster composed of some 1000 stars. As the image here reveals, there is a cloud of interstellar dust surrounding and between the Pleiades and the Solar System, and it is this dust that is responsible for the blue reflection nebulosity. Image courtesy of NASA

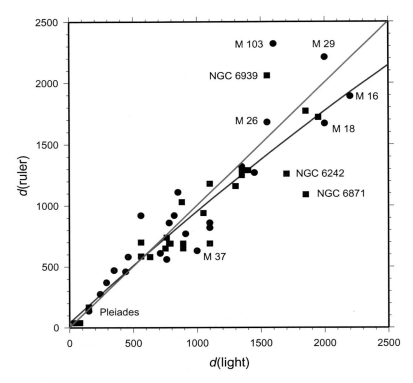

Fig. 3.6 Comparison of the standard ruler and photometric distance measures to a sample of the galactic clusters studied by Robert Trumpler. The *red line* indicates the condition where *d*(ruler)=*d*(light). The *blue curving line* indicates a best fit to the data points and also that typically, for the more remote clusters, *d*(ruler)<*d*(light). Data from *Lick Observatory Bulletin*, Number 420, 1930. A selection of Messier objects are labeled and indicated by *circular points*

The final step in Trumpler's analysis was to construct a graph of *d*(light) versus *d*(ruler). If the data points relating to the sample of clusters produced a straight line of slope 1, then Trumpler knew that this must indicate *d*(ruler)=*d*(light) and that the absorption of starlight by dust is negligible. Figure 3.6 is a plot of the data points Trumpler presented in his 1930 *Publications of the Astronomical Society of the Pacific* research paper, "Absorption of Light in the Galactic System." Some of the galactic clusters fall on the line corresponding to *d*(ruler)=*d*(light), indicating little to no dust and absorption in their line of sight, but others, typically the most distant clusters, fell below the equality line, indicating that *d*(light)>*d*(ruler). Here, at last, was evidence, though not fully convincing, that interstellar dust was all pervasive in the plane of the Milky Way. For the Pleiades, Trumpler found a *d*(ruler) distance of 137 pc, and a *d*(light) distance of 150 pc. This difference in turn indicates that the interstellar dust situated between us and the cluster add (apparently) 13 pc to its distance when gauged by the *d*(light) method.

Problematically, of course, Trumpler's results further indicated that the interstellar dust was not uniformly distributed in the plain of the galaxy, but was clumpy and followed no simple distribution law that could be related to galactic latitude and/or longitude.

In general, however, the amount of dimming, that is, the increase in the d(light) measure caused by interstellar dust decreases at higher galactic latitudes, and it is at its most extreme in the plane of the galaxy, especially so in the direction of the galactic center.

Writing in a leaflet published by the Astronomical Society of the Pacific in July 1932, Trumpler began to harden his arguments. First he noted that if the "dark hole" regions being photographed by Barnard corresponded to a real vacancy of stars, then they would be truly remarkable structures: "… like a narrow straight tunnel some 10,000 to 20,000 light years long… pointed exactly at the solar system." Surely, he reasoned, it was much more likely that the dark nebulae were "largely made up of fine cosmic dust, with which free electrons and atoms, and perhaps larger meteoric bodies also may be mixed." Concerning the meteoric bodies Trumpler is now known to be essentially wrong, but he was certainly right with respect to the dust, free electrons, and atoms.

To bolster his arguments further, however, Trumpler additionally drew attention to the "dark matter"[4] observed in many of the so-called spiral nebulae (recall Fig. 3.3), which "beautifully illustrate how the luminous star clouds of the spiral arms are generously mixed with dark stuff." The picture that was now clearly emerging implicated interstellar dust (small solid particles) as the scattering and absorbing agent of starlight, and that this dust was confined largely to the galactic plane. Indeed, building upon the great survey work of John Herschel, Richard Proctor had drawn attention to an apparent zone of avoidance as early as 1870. This zone stretched around the sky, centered on the plane of the Milky Way, and was devoid of spiral nebula (Fig. 3.7).

The complete dearth of spiral nebulae in this region posed a serious problem to astronomers at that time, and the key questions were why and how. That is, why are there no (spiral) nebulae seen in the plane of the Milky Way and how is the restriction maintained? If nebulae were distributed in space at random then they should be seen in all directions, and there should be no preferential or special orientation with respect to the plane of the Milky Way. In his classic book, *Other Worlds Than Ours* (first published in 1870), Proctor wrote (in his own italics), *"The nebulae seem to withdraw themselves from the neighbourhood of the galaxy,"* adding that, "if this peculiarity is accidental, the coincidence involved is most remarkable." Indeed, Proctor was correct, and an explanation for the apparent sandwiching, above and below, of the starry realm between two layers (or regions) populated mostly by spiral nebulae needed an explanation.

Even as late as 1899, Charles Young, in his classic work *A Text-book of General Astronomy*, noted that the zone of avoidance was largely a complete mystery. "Why the nebula avoid the region thickly starred is not yet clear," writes Young, suggesting, perhaps only partially seriously, that "stars *devour them*, that is, gather in an appropriate surrounding nebulosity so that it disappears from their neighbourhood." In some sense Young was perhaps not too far off the mark with this suggestion, since as will be seen in Chap. 5; stars do in fact form through the gravitational collapse and the appropriation of material previously stored in interstellar gas clouds.

---

[4] Although Trumpler uses the term 'dark matter' he is not invoking the gravitational dark matter of modern-day cosmological interest. Trumpler, like Barnard before him, is simply applying a descriptive expression. It was in 1932–1933, however, that such pioneers as Fritz Zwicky, Jan Oort, and Jocobus Kapteyn first began discussing the 'missing mass' issue with respect to galactic and galaxy dynamics. The mass was not missing, of course, but it is now identified as (the still mysterious) dark matter.

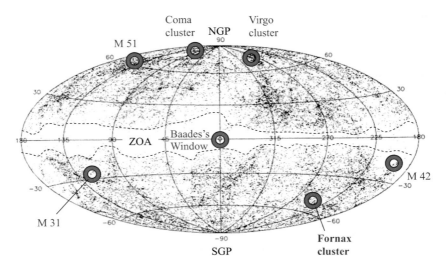

Fig. 3.7   The zone of avoidance is revealed in this all-sky plot of 34,729 galaxies. The image is an Aitoff equal area projection in galactic coordinates, and the zone of avoidance (ZOA) is that region where no galaxy data points are found (the region between the two *dashed lines*). The north and south galactic poles are indicated as NGP and SGP, respectively, and the center of the diagram indicates the location of galactic center and Baade's Window (see later). Background galaxy data plot from N. A. Sharp, 986, PASP. **98**, 740

The zone of avoidance, we now know, through the pioneering work of Barnard and Trumpler, is entirely the result of interstellar dust, and predominantly dust within the disk of our galaxy, fully absorbing and scattering from our line of sight the faint light from distant galaxies. The galaxies are there, but we cannot see them at optical wavelengths. This very same result is illustrated in our earlier star and nebula count data (shown in Fig. 1.19a, b), where it can be seen that by far the greater number of the galaxy counts are registered in those selected regions situated close to the galactic poles. It is within these latter two directions that we are actually looking out of the disk of the Milky Way Galaxy, rather than into the plane of the disk, and accordingly the dust distribution is at its lowest concentration in our line of sight.

That high dust concentration regions, dark nebula, are most probably associated with star formation was first articulated by Bart Bok and Edith Reilly (both of Harvard College Observatory) in a brief but important paper published in the *Astrophysical Journal* in 1947. Bok and Reilly concentrated their discussion and study to the smaller dark nebula, and proposed such nebula might reasonably be called globules. The globules typically have masses between a few and 10 times that of the Sun and characteristic dimensions smaller than a light year (Fig. 3.8). Building upon star formation ideas presented by Lyman Spitzer and Fred Whipple in the early 1940s, Bok and Reilly argue that the small dark clouds (now, with great disservice to Reilly, generally called Bok globules), "probably represent the evolutionary stage just preceding the formation of a star." Bok and Reilly were indeed correct in their suggestion, but we shall reserve details of the process until Chap. 5.

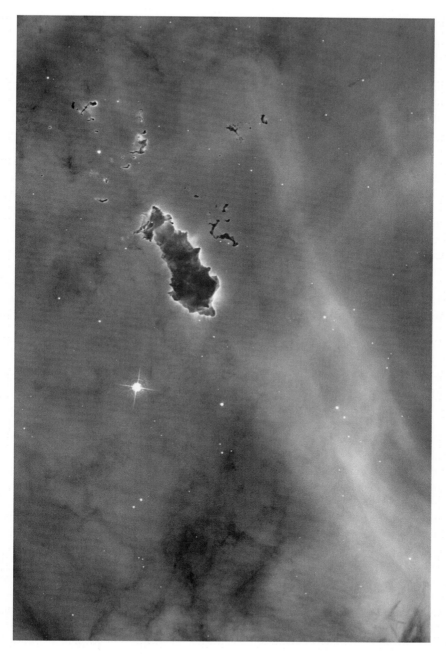

Fig. 3.8 Bok globules in the Carina Nebula. The large globule, towards image center, has been appropriately nicknamed the 'caterpillar,' and it is indeed a region in which new stars are beginning to form. The glowing edges that highlight the globules are produced by a photoionization effect caused by the ultraviolet radiation of other stars located within the Carina Nebula. The nebula is some 7500 light years distant from the Sun. Image courtesy of NASA/HST

Baade's Improbable Window

One line of argument that Trumpler used to reject the idea that the 'dark-holes' in the sky were regions truly void of stars was that it would be a truly remarkable geometrical arrangement. The void would have to be a tube-like exclusion zone, stretching through the entire length of the universe, and Earth would have to be positioned so as to see exactly, or nearly so, down the length of the tube. Indeed, such a situation would be an unlikely circumstance, but as revealed earlier, the dark-holes are not voids but localized regions of high interstellar dust concentration.

Although interstellar dust fills the disk of our galaxy it does not do so uniformly. Dust clouds can be clumpy, and they can be drawn out into long ribbons; there are regions of very high dust concentration (Barnard's black-holes, Trumpler's dark matter, and Bok's globules) and regions of very low dust density. Given such a picture of the dust distribution within the disk of our galaxy (recall, as well, the edge-on galaxy in Fig. 3.3) it is remarkable that there are, in fact, several narrow windows, or pathways of view, through which the great depth of space leading to the galactic center can be viewed at optical wavelengths. By far the best known and best studied of the very low dust absorption corridors, and some six are now known, in the direction of the galactic center is Baade's Window.

Walter Baade announced the discovery of his unlikely window, with its view along the relatively dust-free tunnel to the galactic center, at the 1946 Reno, Nevada, meeting of the Astronomical Society of the Pacific. At this particular time Baade was in search of stars located within the galactic bulge, and to find them he took a leaf right out of William Herschel's observing book. Baade reasoned that the line of sight along which the stars would be seen must be at least 25,000 or more light years in length, as reasoned by Harlow Shapley's analysis of the distribution of globular clusters. Accordingly, if such distant stars were to be visible there would also be a large number of foreground stars, and observationally this would correspond to a relatively bright patch of sky with a fairly uniform and very high density of stars per unit area. Baade's task, therefore, was to seek out those regions of brightest starlight in the direction of Sagittarius—the region implicated by the observed globular cluster distribution (recall Fig. 1.1c) as containing the galactic center.

And, sure enough, in a small 1-degree diameter region in the sky located close to the star γ Sagittarius, he found his window (Fig. 3.9). Indeed, the window was centered almost exactly on globular cluster NGC 6522—a cluster, in fact, that William Herschel had discovered some 160 years earlier, on June 24, 1784. The number of stars visible through Baade's Window is truly staggering, and it is now recognized as one of the very few regions on the sky through which stars located within the galactic bulge are visible.

The view through Baade's Window does not reveal the true galactic center, but rather it shows those stars located in a region some 450 light years from it. By studying the brightness variations of some 50 RR Lyrae variables that had been identified in his photographic survey plates, Baade estimated that the Sun was located some 27,000 light years away from the center of the galaxy (now estimated to be more like 26,000 light years). While Baade reaffirmed the fact that the central regions of the Milky Way Galaxy are certainly remote, he showed, for the first time, that in a very small window on the sky, the stars close to central core were amenable to study.

Baade's interest in observing the galactic center evolved from his earlier studies of the Andromeda Galaxy (M31; recall Fig. 3.2). Indeed, in 1944 Baade had introduced the

Fig. 3.9 Baade's Window. The high and uniform density of stars visible in this image imme-diately betrays the great depth to which they are being seen. NGC 6522 is to the *upper right* in the image, while a second globular cluster, NGC 6528, is visible to the *lower left*. Image by Adam Block, Mount Lemmon SkyCenter, University of Arizona

concept, still held to this day, that in a very broad sense stars can be divided into two main groups based on composition and dynamics. These groups are designated Population I and Population II. The scheme is specifically based on measuring the spectral characteristics, orbital motion, and abundance of heavy elements (see later) of numerous stars.

Baade placed those stars that were metal poor compared to the Sun, and moved in long elliptical orbits inclined to the galactic disk, in the Population II category. In contrast, stars with metallicity similar to or even greater than that of the Sun, moving along near circular orbits within the galactic plane, were designated Population I objects.[5] Baade developed his population scheme in response to the observed differences between those stars observed

---

[5] In the late 1970s an additional Population III group was added to Baade's scheme. This population is reserved for the very first formed stars, which have zero heavy elements and are composed entirely of hydrogen and helium. No such population has as of yet been unambiguously observed, but their detection is one of the cornerstone projects of the James Webb Space Telescope program, due for launch in 2018. See Chap. 5 for further details.

in the central bulge of the Andromeda Galaxy (Population II stars) and those that were observed in its disk (Population I). Accordingly, Baade reasoned that if Population II stars could be observed in the central regions of the Milky Way, this would indicate that our galaxy must also be a spiral galaxy—indeed, it must be something like M31.

Baade was, it turns out, partially right in his reasoning, and while Population II stars are detected in the core regions of our galaxy, so, too, are Population I stars. The population designation is an informal proxy for formation age (see Chap. 5), the notion being that the oldest stars (extreme Population II) will have formed in environments containing fewer heavy elements such as when the galaxy itself was much younger. These stars, for example, are found within globular clusters and the galactic halo population. In contrast, the most recently formed stars, extreme Population I objects, will have formed from those regions of the interstellar medium with high metallicity. Such stars, in fact, constitute the young, hot, highly luminous, and short-lived O spectral type stars that delineate the spiral arm structures of our galaxy.

Having found a window opening onto a grand view, then there is nothing better than taking a long and detailed stare at what the framed vista reveals. It is exactly this approach that has been adopted by OGLE, and through its near catatonic gaze at those stars located within Baade's Window, the project has discovered some of the most distant exoplanet systems known. OGLE is the acronym for the Optical Gravitational Lensing Experiment, which has its survey headquarters at the University of Warsaw in Poland.

As the acronym indicates (when unraveled) the observational technique at the heart of the OGLE is that of gravitational lensing. This particular observing technique exploits the effect of the general relativistic bending of light produced by a massive object. In the case of so-called microlensing the lensing object is a star located between the observer and another much more distant, background star. Important for the observational technique, when the foreground and background star approach near exact alignment, the gravitational lensing effect results in a distinct increase in the brightness of the background star (Fig. 3.10). This is what the OGLE survey is looking for. By repeatedly scanning the same region of the sky, the researchers are hoping to record the characteristic rise and then fall in the brightness of a distant background star, due to the passage of a foreground star across the observational line of sight.

The chances of recording a lensing event are greatly heightened in those regions in which the number of background stars per unit area in the sky is very high, and this naturally favors survey regions such as Baade's Window. The OGLE began in 1992, and it has now recorded numerous gravitational lensing events. The international consortium of astronomers engaged in the OGLE program has found among these events some of the most remote exoplanetary systems within our galaxy (Fig. 3.11). The present distance record holder has the somewhat ungainly (but data-packed) title of OGLE-2005-BLG-390Lb. This particular planet has an estimated mass some 5.5 times that of Earth, and it orbits its parent dwarf star (with a mass about 1/5 that of the Sun) at a distance of about 3 AU. The lensing event was first detected by the OGLE observers in August of 2005, and subsequently followed by various research groups for several months. Although the planet itself is not so extraordinary, what is remarkable is that its parent star is located some 21,500 light years from the Solar System. At this distance from us, the star and planet will have a bedazzling, birds-eye view of the galactic bulge and center. The night sky view from this planet must surely be a truly spectacular sight.

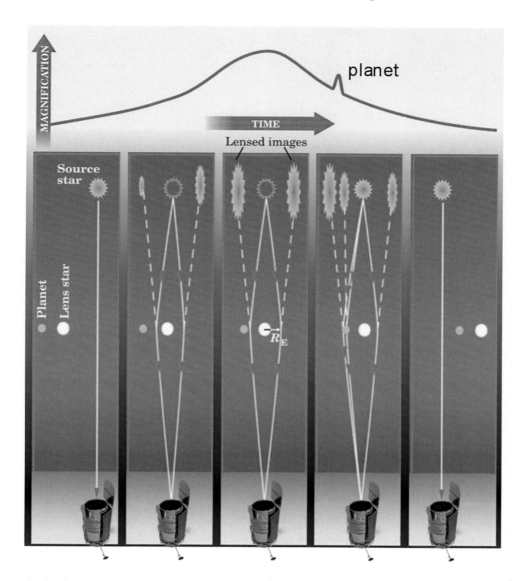

Fig. 3.10 A schematic time sequence series of images showing the development of a gravitational lensing event. The lens star moves slowly across the observational line of sight to a distant background, source star. As the approach distance narrows, so the brightness of the background star increases (as shown by the magnification curve at the *top* of the diagram). If the foreground lensing star chances to have an orbiting planet in the right geometrical arrangement, then it may produce a small duration brightness spike (see Panel 4). Image courtesy of NASA Goddard Space Flight Center

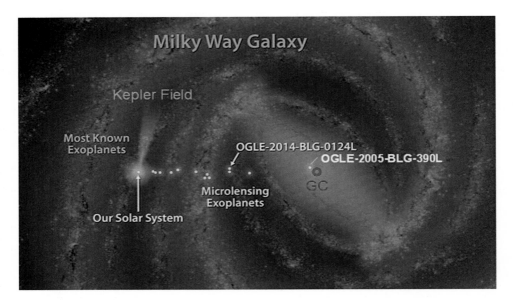

Fig. 3.11 Stretching out like pearls on a necklace, this image reveals the exoplanetary systems found by the OGLE survey in the direction of Baade's Window. OGLE-2014-BLG-0124L is located some 13,000 light years from the Sun, and consists of a half-Jupiter mass exoplanet orbiting its parent star at a distance of about 3 AU. Even deeper along the field of view is OGLE-2005-BLG-390L, located within the galactic bulge region. By far the greatest numbers of known exoplanetary systems (discovered to date) have been found with the Kepler satellite (see Chap. 6). During its first phase of operation Kepler used a point and stare mode to search for planetary transits in the rich star fields located in the direction of Cygnus. Background image courtesy of NASA/JPL

Seeing the Blues

Take another look at Fig. 3.5, which shows the Pleiades cluster (M45). What is remarkable about the image is not so much that its shows a grouping of young, hot stars (recall, however, the discussion in Chap. 1), but that the image appears so oddly smudged and blue in color.

At first glance it might seem that the image has been badly handled, with the printer's inky fingerprints carelessly left as a blemish on the original print, but this is most definitely not the case. What the image of the Pleiades shows is the direct presence of interstellar dust and the blue coloration provides important information about the physical size of the interstellar dust grains. Indeed, M45 shows the characteristics of what are known as reflection nebula.

The idea that the stars of the Pleiades cluster might be associated with a cloud of interstellar dust was first discussed by Vesto Slipher in *Lowell Observatory Bulletin* (# 55) published in 1912. Slipher drew this conclusion from the observation that the nebulosity produced the same absorption line spectrum as that associated with the cluster stars, and this in turn implicated the reflection (scattering) of starlight as the reason for the nebulosity.

In the case of the Pleiades, Slipher's conclusion was entirely correct, although as an indicator of the uncertainty of the times, he further went on to suggest that perhaps spiral nebula might also be explained in the same manner, with a central star "enveloped and beclouded by fragmentary and disintegrated matter which shines by light supplied by the central sun."

What we are seeing in the image of the Pleiades is an effect we actually witness every sunny day—the blue sky above our heads. Indeed, it took no less than the major intellect of Albert Einstein to show that the sky should indeed be blue. The details are hidden away, however, in a lengthy and complex paper entitled "The Theory of the Opalescence of Homogeneous Fluids and Liquid Mixtures near the Critical State," published in the journal *Annalen der Physik* in 1910.

Einstein's early research, since the time of his 1905 doctoral thesis, was concerned with characterizing the size of molecules and understanding Brownian motion. In his 1910 paper, however, he looked at the way in which molecules might interact with electromagnetic radiation. What Einstein (and others at the time) wanted to determine was how light was scattered, that is, how its path is changed or its energy absorbed, as it moves through a gas or a fluid. Indeed, we now know that molecules can scatter light because the electromagnetic field of the incident light wave induces an oscillating electric dipole movement within the molecule.[6] And this, in effect, acts to re-radiate a light wave, identical to the incident one but in a new direction of propagation. That is, the light wave is effectively scattered. If all wavelengths of light were scattered equally then starlight, as it propagated through space, would simply become fainter and fainter with increasing distance (this effect being in addition to the inverse square law of light propagation).

In essence, the more scattering there is along the line of sight to a star, the less starlight there is that will actually get to the telescope detector. Nature, however, has made light scattering wavelength dependent, and this makes the life of the astronomer somewhat more problematic (but also more interesting).

The first experimental indication that we should indeed see a blue sky was proffered by Irish physicist John Tyndall in 1859. Working in the laboratory of the Royal Institution of Great Britain in London, Tyndall discovered that a clear liquid (or gas) holding small particles in suspension (a so-called colloid) scattered shorter blue wavelengths of light more efficiently than those of longer wavelengths closer to red light. This so-called Tyndall effect can be easily demonstrated with a modern-day laser pointer and a clear glass beaker of water. Partially fill the glass beaker with tap water, and place it on a piece of white paper. Shine the laser pointer light through the side of the glass and look down on the liquid. What will you see? Well, at this stage nothing much. The light beam enters and then exits the sides of the beaker, but there is no trace of the beam as it crosses through the water in between.

Now, one small drop at a time, add some milk to the water solution. At first not much will happen, but as the concentration of triglyceride globules within the water-milk mixture increases, the laser beam will gradually become visible. What is happening at this stage is that the milkfat globules are beginning to scatter more and more of the light from the laser beam, thereby making the physical path of the laser beam through the water visible. This is the Tyndall effect at work.

---

[6] An electric dipole is formed when the positive and negative charges within a molecule (or atom) take on a non-uniform distribution.

The wavelength dependence of the light scattering in Tyndall's experiment was later investigated theoretically by British physicist John Strutt (later Lord Rayleigh) in 1871, and he was able to show that when the particles of radius a, responsible for scattering, are smaller than the wavelengths of light being considered, $a < \lambda$, then the intensity of scattering, I, varies as the inverse fourth power of the wavelength: $I \propto 1/\lambda^4$. Tyndall and Rayleigh both realized that this scattering effect, as seen in experiments on colloids, could explain why the sky is blue. They argued, therefore, that as sunlight moved through the atmosphere it would encounter very small dust particles and water vapor droplets, and these would act as scattering nuclei.

Also, since the wavelength of red light is almost twice that of blue light, the intensity of scattering would be $2^4 = 16$ times greater for blue light than red, and hence the sky will appear predominantly blue.

There is a problem with this explanation, however. Since the total amount of water vapor droplets and dust in the atmosphere in any one area is highly variable, the total amount of scattering should also vary. That is, on some days the sky should not be as blue in color as it is on others. This is where Einstein enters into the story, since in 1911 he showed that molecules will also scatter light, and scatter light in the same characteristic manner as Strutt (Lord Rayleigh) had deduced for solid particles in suspension.

The sky is blue, therefore, because of Rayleigh scattering and the fact that the atmosphere is largely composed of nitrogen (N_2) and oxygen (O_2) molecules. Any additional dust and aerosol mixture in the lower atmosphere will additionally enhance the scattering effect, especially when the Sun is low on the horizon, and this is the reason for the spectacular orange sunsets that are often seen over highly polluted city skylines. The very same effect happens in interstellar space. As a direct consequence of the size distribution of interstellar dust grains (see later), the shorter blue wavelengths of starlight are scattered more efficiently than the longer red wavelengths of starlight, and this results in what astronomers call interstellar reddening (Fig. 3.12). The preferential scattering of blue light also results in the spectacular appearance of reflection nebula, such as that seen around the Pleiades (Fig. 3.5) and, for another particularly beautiful example, the reflection nebula IC 2631 (Fig. 3.13).

The size of interstellar dust grains, given that they preferentially scatter blue light, indicates that they must have a size smaller than, but comparable to, the wavelength of blue light, which indicates sizes $a \sim 3 \times 10^{-7}$ m. It turns out, however, that the number density $n(a)$ of dust grains in the interstellar medium decreases steeply with increasing size, with $n(a) \sim a^{-3.5}$. This results in extinction being strongly biased towards shorter wavelengths of light, and it also results in a non-Rayleigh scattering extinction law.

Indeed, in a landmark paper published in 1930, Robert Trumpler showed that the amount of interstellar extinction A (to be described shortly) does vary with wavelength, but with $A \propto 1/\lambda$. Figure 3.14 shows a modern-day plot of the wavelength dependency of interstellar absorption as derived for the Milky Way Galaxy and the Small and Large Magellanic Clouds. The figure reveals a number of mysterious and important points about extinction. Firstly, the absorption is not strictly what's expected from pure Rayleigh scattering, and this is a direct consequence of the different sizes exhibited by interstellar dust grains.

Nobel Prize-winning physicist Edward Purcell actually showed in a famous research paper published in 1969 that grains with a size comparable to the wavelength of light being scattered do so in a manner that varies as the inverse wavelength. Furthermore, Fig. 3.14 reveals that the extinction at a given wavelength varies from one sample region

Fig. 3.12 This image shows the Horsehead Nebula—a Bok globule located within the Orion Nebula (M42). A close examination of the stars seen near to, and slightly within, the boundary of the nebula reveals that they are highly reddened. This is a direct consequence of the dust within the cloud preferentially scattering blue light away from the line of sight. The blue background to the silhouetted image is a direct consequence of blue light being scattered by dust into the observational line of sight. Image courtesy of NASA/HST

to another. The extinction at shorter, UV wavelengths of light, for example, is far higher in the Small Magellanic Cloud (SMC) than it is in either the Large Magellanic Cloud (LMC) or the Milky Way Galaxy. This indicates a greater predominance of smaller-sized dust grains within the SMC than that encountered within the LMC and the Milky Way Galaxy, and the reason for this is presently unclear.

Figure 3.14 also reveals that for visual and near-UV wavelengths of light the extinction varies, as found by Trumpler, according to the inverse of the wavelength—that is, it increases as the wavelength decreases. Roughly the same inverse wavelength law applies in the far UV region of the spectrum as well [3]. There is, however, a distinct bump in the LMC and Milky Way extinction curves at a wavelength $\lambda = 2.175 \times 10^{-7}$ m (2175 Å = 4.6 μm^{-1}). The origin for this bump, first noted in observations made in the 1960s, is not entirely clear, but it may possibly relate to a dust composition feature, related to the presence of organic carbon or amorphous silicate inclusions. Why the bump feature does not appear in the SMC extinction curve is presently unclear, but it may be related to its presently observed high star formation rate and its apparently rapid recycling of interstellar gas and dust.

Fig. 3.13 Reflection nebula IC 2631 in the Chamaeleon star formation complex. The nebula is illuminated by the young, hot, high luminosity T Tauri star HD 97300. The host star, surrounding dust cloud, and nebula are located some 500 light years from the Sun, in the direction of the southern hemisphere constellation of Chamaeleon. Image courtesy of the European Southern Observatory

In general it is clear that interstellar grains are predominantly made of amorphous silicates along with add-on mixtures of carbonaceous inclusions, such as graphite and carborundum. The origin of the interstellar grains will be discussed in more detail in Chap. 5, but it is clear that they form in the outer envelopes of carbon-rich giant stars during their late stage evolution and within the ejected gas envelopes produced by novae and supernovae explosions.

From the curves shown in Fig. 3.14 it is apparent that some of the grains in the interstellar medium must have sizes of order 0.1 μm in order to account for the visual extinction, while others must be as small as 0.01 μm to account for the far-UV extinction. Even smaller interstellar grains must exist, and have indeed been found as nanodiamonds and titanium-carbide nanocrystals within carbonaceous chondrite meteorites discovered on

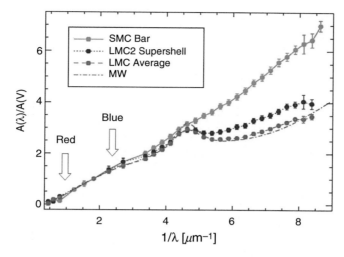

Fig. 3.14 Interstellar extinction curves for the Milky Way and selected regions of the SMC and LMC. The *vertical arrows* mark the locations for the wavelengths corresponding to blue and red light. Note that the x-axis is plotted in terms of the inverse wavelength

Earth (Fig. 3.15). It is through the analysis of the various oxygen and carbon isotope ratios that such inclusions have been identified as pre-solar in composition—that is, the grains must have formed in regions other than that of the young Sun.

Remarkably, large interstellar grains, with sizes in excess of 1 μm, have been directly sampled by dust-detection instruments carried aboard the Ulysses and Galileo spacecraft.[7] Even larger interstellar grains, with sizes between 10 and 100 μm, have additionally been detected with sensitive all-sky radar systems as they are destroyed, by ablation, in Earth's upper atmosphere. The Meteor Physics Group at Western University in Ontario, Canada, under the direction of Peter Brown, has been particularly concerned with the study of interstellar meteors using the Canadian Meteor Orbit Radar (CMOR).

Interstellar meteors are specifically identified by their speed of encounter that, given their non-solar origin, will be greater than some 75 km/s. A 2005 report by Robert Weryk and Peter Brown found, for example, that of the more than 1.5 million meteoroid orbits recorded in a 2.5-year study interval, some 0.0008 %, that is, about 10 meteors, were of interstellar origin. In addition to the radar studies, interstellar grains have also been directly sampled by high-flying aircraft and balloon-borne experiments in Earth's upper atmosphere.

That many interstellar dust grains must be non-spherical in shape (as indicated by Fig. 3.15) is betrayed by the way in which starlight, in many directions of the Milky Way, is polarized—that is, the light wave oscillates in a preferred (rather than random) direction. Such an effect will only come about if the interstellar grains are elongated and substantially aligned over a large volume of space. The mechanism that controls grain alignment is now recognized to

---

[7] The Ulysses spacecraft mission (1990–2009) was specifically concerned with the study of the Sun's atmosphere and the solar wind, while the Galileo spacecraft (1989–2003) conducted a highly successful mission to Jupiter.

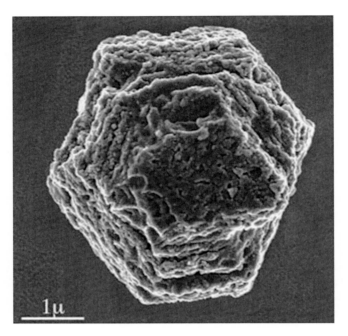

Fig. 3.15   Electron microscope image of a pre-solar silicon carbide (SiC) dust particle sampled from the Murchison carbonaceous chondrite meteorite. This grain, which is some 5 μm in size, formed in the outer envelope of an aging star long before being incorporated into the gas and dust cloud that produced our Solar System. Image courtesy of *The Source*, December 12, 2011, Washington University in St. Louis

be the Milky Way's magnetic field. Indeed, careful mapping of the polarization directions has enabled astronomers using data gathered with the Planck satellite[8] to map out the twists and turns of our galaxy's magnetic field (Fig. 3.16). The importance of the galactic magnetic field and especially its role in moderating star formation will be discussed in Chap. 5.

In the process of absorbing starlight, interstellar grains will become heated, and this heat will be re-radiated back into space in the form of infrared radiation. Indeed, the characteristic temperatures of isolated interstellar dust clouds are found to be between 20 and 100 K. Figure 3.17 shows an all-sky infrared map based on observations obtained with the COBE satellite. The map is orientated along the galactic plane (the Milky Way), and from the high infrared emission detected in this plane it is evident that most of the interstellar dust resides in the disk of the galaxy. Indeed, the dust is primarily confined to within a zone with an effective thickness of about 200 pc on the galactic plane.

Most of the interstellar dust resides interior to the Sun's orbit around the galactic center, and it is evident from the emission intensity that the dust it is not uniformly distributed but is rather arranged in distinct high-density clouds. Running around the edge and across the

---

[8] The primary task of the Planck spacecraft was to map out anisotropies in the cosmic microwave background radiation. Before such a map could be constructed, however, the effects of the thermal emission due to galactic dust had to be removed.

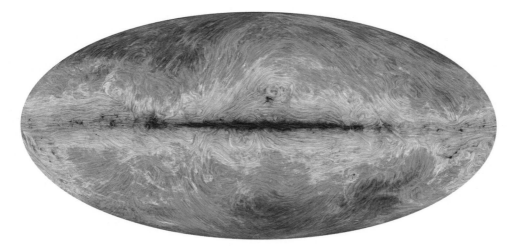

Fig. 3.16 Galactic magnetic field orientation, as mapped out by the dust-induced polarization of starlight. The data was gathered during the Planck satellite mission (2009–2013). The color scale in the image indicates the total intensity of dust emission, revealing that the dust is predominantly located in the galactic plane. The oil painting-like texturing on the image is based on the deduced direction of polarization, which in turn shows the orientation of the galactic magnetic field. Image courtesy of the European Space Agency

Fig. 3.17 COBE satellite all-sky map of the infrared sky at a wavelength of 60 μm. The image has been arranged according to galactic coordinates, with the band of the Milky Way running horizontally across the central diameter. The image is color coded according to emission intensity, with the strongest emission regions being shown in white. To the *far right* and below the galactic plane is the Orion Nebula (M42). To the *lower right* of center the two isolated bright regions are the Small and Large Magellanic Clouds. The Andromeda Galaxy (M31) is also just visible below the galactic plane

central diagonal from left to right is a faint emission glow (given a blue coloration in the image). This is the thermal emission due to dust grains located within the ecliptic plane of the Solar System. However, this latter dust distribution is the result of asteroid upon asteroid collisions and the out-gassing of cometary nuclei within the Solar System itself.

Interstellar Extinction

The effect of interstellar dust on starlight can be considered a pure absorption process, and it will accordingly vary with respect to the size and number of grains Ng in the line of sight. Schematically, the dust extinction can be modeled as shown in Fig. 3.18, and importantly, we can estimate the number of dust grains Ng in the line of sight by a star count method. It's back to $1+1+1=3$ again.

The trick to determining the amount of interstellar extinction in a specific direction is to compare the number of stars counted, over a range of magnitudes, in a field direction (assuming that there is no dust in the line of sight) against the number of stars counted, over the same magnitude range, in the direction of interest. The key point here is that the number of stars at a given magnitude in the direction of a dust cloud should increase in the same manner as the field star count (recall Chap. 1 and see the Appendix in this book), but the relative number of stars in the direction of the dust cloud will be reduced because of the intervening dust.

Figure 3.19 shows a black and white negative image of a small region of the Coalsack Nebula (recall also Fig. 1.5), and superimposed on the image are two equal area circles. Our task now is to count the number of stars visible in each circle; those stars contained with the central circle will give the nebula count N, while those stars contained within the circle to the lower right of the image will give the field count N_0. Fifteen minutes of careful counting by the author's eye reveals that (in round numbers) $N_0=1800$ and $N=210$. These number counts indicate a visual extinction of about 2 magnitudes between our two sample regions in the Coalsack Nebula. This amount of extinction indicates that the total number of grains per square meter in the sky in the direction of the Coalsack Nebula is $Ng \approx 3 \times 10^{12}$, and, assuming that the Nebula is 600 light years away, the total mass of dust grains in the nebula region under study is about 4.5×10^{28} kg, or some 1/22 the mass of the Sun.

In general it is typically convenient to express the extinction in terms of the ratio A/L, with the extinction being described in magnitudes per kiloparsec. In the solar neighborhood, for example, the observations indicate an average extinction at visual (V) wavelengths [3] of some $<A_V/L>=1.8$ mag/kpc (see Fig. 3.14 for $A(\lambda)/A(V)=1$). Taking a typical dust grain to have a radius of 0.1 μm, the number density of dust grains in the solar neighborhood amounts to $n_{dust}=10^{-6}$ m^{-3}; this corresponds to the typical grain separation

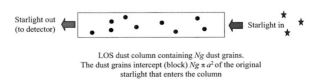

LOS dust column containing Ng dust grains.
The dust grains intercept (block) $Ng \, \pi \, a^2$ of the original
starlight that enters the column

Fig. 3.18 Schematic diagram of the direct absorption of starlight by interstellar grains

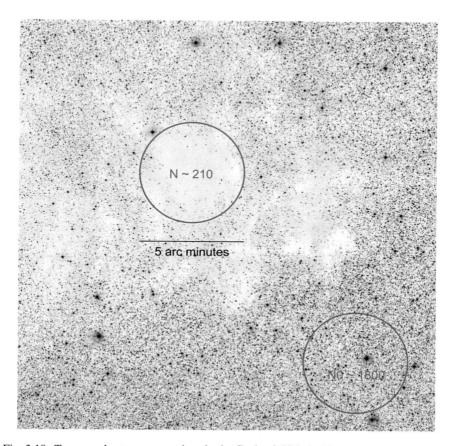

Fig. 3.19 Two sample star count regions in the Coalsack Nebula (shown in a negative black and white image format). Each circle has an approximate diameter of 5 minutes of arc. One circle (*to lower right*) is set over a background region in which interstellar dust should have a minimal effect upon the star count numbers. The other circle is situated over a heavily obscured region in the nebula. Background star field and nebula image courtesy of the European Southern Observatory

being of order 100 m.[9] With the individual grains having a density of 2500 kg/m^3, the mass density of interstellar dust in the solar neighborhood evaluates to $\rho_{dust} = 10^{-23}$ kg/m^3, which on the grander galactic scale translates to about 10^{-4} solar masses of interstellar dust grains per cubic parsec of space. Taking the latter mass density to be a typical value, then the total amount of matter in the form of interstellar dust within our galaxy adds up to about 3×10^7 solar masses.[10]

---

[9] It is remarkable to think that the typical separation to size ratio for interstellar grains is of order 1 billion. Even stars are more closely packed than this. The separation to size ratio for Proxima Centauri and the Sun, for example, is a relatively small 29 million.

[10] Here we take the galactic disk to have a radius of 10 kpc and a depth of 1 kpc.

Where there is dust there is gas. This is the modern mantra with respect to the interstellar medium. In Chap. 2 we saw that the presence of hydrogen within the interstellar medium is clearly implicated through the red light Hα recombination line observed in emission nebula. What, however, is the relative concentration of the gas compared to the dust?

When gas is in emission it is a relatively straightforward task to map out its distribution and determine its overall abundance; it is more problematic, however, to map out the distribution of a low temperature gas. As we shall see in Chap. 4 the hydrogen gas component of the interstellar medium exists in two forms, neutral hydrogen HI and molecular hydrogen H_2. Far removed from the ionizing influence of hot stars the HI atoms will be in their $n=1$ ground state (recall Fig. 2.11); the H_2 molecule will likewise be in a low energy state.

Neither the cool hydrogen nor the H_2 molecules are directly observable at optical wavelengths, but their presence in the line of sight to a star can be deduced through observations at ultraviolet wavelengths. By moving to the ultraviolet part of the electromagnetic spectrum the existence of hydrogen can be inferred by looking for Lyman alpha (Lα) absorption lines in the spectra of hot stars. The Lyman alpha absorption line involves the electron transition from the $n=1$ ground state to the $n=2$ first excited state of hydrogen, and these transitions are enabled by ultraviolet wavelengths photons.

The trick to finding neutral hydrogen, therefore, is to observe the spectrum of high temperature stars, which emit copious amounts of UV photons. The number of HI atoms, $N(HI)$, in the line of sight can then be determined by measuring the strength of the Lα absorption line. In this manner, the bright, high temperature background stars act like spotlights that illuminate the amount of neutral hydrogen in the astronomer's line of sight. Likewise, the number of H_2 molecules, $N(H_2)$, in the line of sight can be inferred by the observation of the allowed molecular rotation vibration transition lines[11] (see Chap. 4 for details on molecular line transitions). The upshot of the UV observations, therefore, is that the total amount of hydrogen $N(H)=N(HI)+2N(H_2)$ can be determined in the line of sight to various bright, high temperature (e.g., O, B spectral type) stars. Furthermore, the same search-light stars used in the determination of the line of sight $N(H)$ values can be studied to reveal the amount of dust extinction $A(V)$ in the same direction. These two measures, it turns out, are reasonably well correlated, with $N(H)/A(V) \approx 2 \times 10^{17}$ atoms per square meter per magnitude of extinction.

With such observational measures in hand the dust to gas density can be estimated, and typically it is found that $\rho_{dust}/\rho_H \approx 0.01$, with the number density of hydrogen, in the solar neighborhood, being $n_H \approx 10^6$ m^{-3}. We are now in a position to characterize the hydrogen component of the interstellar medium, finding that the typical spacing between neighboring hydrogen atoms is of order 1 cm (rather than the typical 100 m spacing of dust grains) and that $\rho_{gas} \approx 10^{-21}$ kg/m^3. The gas in the interstellar medium, therefore, is much more abundant than the dust and indeed following our earlier calculation for the dust mass within our galaxy that for the hydrogen gas adds up to about 4×10^9 solar masses. This gas and associated dust, of course, is the galaxy's storehouse of raw material for making new stars, and the conversion process by which the interstellar medium is transformed into stars will be discussed in Chap. 5.

The Bok globule Barnard 68 (Fig. 3.20) offers a dramatic example of very high dust extinction at optical wavelengths. It is a remarkable structure estimated to be some 0.16 light years across. Indeed, transported to our Solar System, Barnard 68 would sit comfortable within the region interior to the inner edge of the Oort Cloud. Not one star exists between us

---

[11] In the UV part of the spectrum these are known as the Lyman-Werner transition bands.

Fig. 3.20 Barnard 68, a Bok globule located some 400 light years away. At visual wave-lengths no background starlight can penetrate the nebula, resulting in its entirely dark, silhou-etted appearance. Image courtesy of ESO

and the nebula to pockmark its dark surface, and the extinction due to the globule, at visual wavelengths of light, amounts to an estimated 25 magnitudes. It is an oppressively dark portal. The amount of gas and dust contained within Barnard 68 is estimated to be about twice that of the Sun, and it would appear from the observations (to be discussed in Chap. 5) that the cloud is teetering on the brink of gravitational collapse to form a star.

As a remarkable demonstration of the wavelength dependency of extinction, however, Fig. 3.21 shows a series of images of the globule taken in the infrared part of the spectrum. The effect is astounding, and as one moves to longer and longer wavelengths the dark nebula slowly disappears, with the background stars accordingly beginning to emerge. Indeed, at $\lambda = 2.16$ μm, the nebula is barely noticeable. The ability to peer through dust and dust clouds at longer wavelengths of light has, in the modern era, been vital to the discovery and understanding of galactic structure. With infrared and radio telescopes

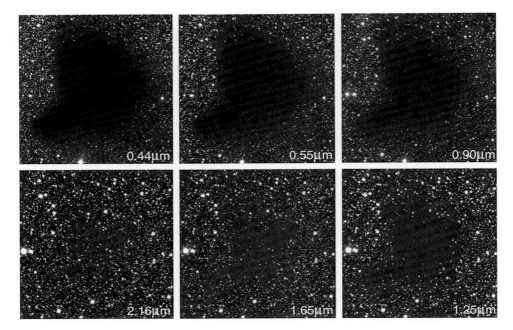

Fig. 3.21   Barnard 68 imaged at a selection of infrared wavelengths between 0.44 to 2.16 μm. The decrease in extinction with increasing wavelength is highlighted between the *top and bottom leftmost* images. Image courtesy of ESO

astronomers have been able to circumvent the optical restrictions of the centrally offset Barnard's Window to view the very heart of our galaxy.

Located some 8000 pc away, the strong radio source Sagittarius A* identifies the location not only of the galactic center, the very hub of the Milky Way, but the lair of a massive black hole. Entirely invisible at optical wavelengths, there being something like 14 magnitudes of extinction between the Sun and the galactic center, the galactic core is a violent and active place. Radio telescope observations reveal numerous supernova remnants, star formation regions, and massive clouds of superhot, ionized gas. Furthermore, observations at infrared wavelengths have revealed the motion of stars located right at the galactic core. Indeed, the observed elliptical orbit of one star, somewhat unceremoniously known as S2, reveals that the central black hole must weigh in at something like 4.1 million solar masses [4].

Barnard 68 affords us a glimpse ahead to the topics to be covered in Chaps. 4 and 5. Indeed, Barnard 68 is a molecular cloud core. While not yet a protostar it is the crepuscular body, spread large in space, of a star yet to be. Now that the material to make a star has been assembled, all that is required next is for the globule's self-gravity to win over the internal mechanisms of pressure support. We wait then with baited breath, since once self-gravity successfully sinks its teeth into the cloud, so a star will form within a few ten thousands of years.

The details of the cloud collapse phase and the process of new star formation will be followed in Chap. 5. For the moment, however, we continue to concentrate on Barnard 68. Indeed, Fig. 3.22 continues the process of probing the globule at different wavelengths of light. In this figure we see new and important results—the top row of Fig. 3.22 shows the

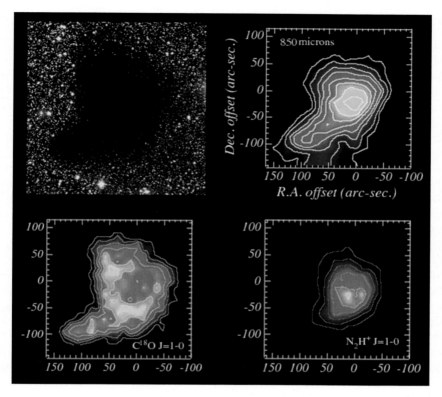

Fig. 3.22 *Top row:* Images of Barnard 68 taken in the visual, and in the far infrared (at 850 μm). *Bottom row:* Images of Barnard 68 made at wavelengths of 2.73 and 3.22 mm, corresponding to emission lines associated with the $C^{18}O$ and N_2H^+ molecules. Image from https://inspirehep.net/record/1254550/plots

globule at optical wavelengths and in the infrared, while the lower two images have been constructed from millimeter wavelength observations of the rotational emission lines (see next chapter) associated with the carbon monoxide $C^{18}O$ molecule and the dinitrogen monohydride cation N_2H^+. Such observations reveal Barnard 68 to be a bona fide molecular cloud, and they further reveal that the molecules are largely clumped towards the central regions of the globule, where the dust grain number density is at its highest and the temperature is at its lowest.

As will be discussed more fully in the following chapter, the dust is vital not only for the formation of the molecules themselves but also for keeping the central regions of the globule cool and shielded from disruptive stellar radiation. Indeed, a detailed analysis of Barnard 68 by João Alves (European Southern Observatory) and co-workers in 2001 found that the cloud temperature is a frosty 16 K and that the center to edge dust density ratio is a whopping 16.5, these latter numbers indicating that the typical number density of grains at the center of Barnard 68 is of order 2×10^{-5} m^{-3}, with the typical grain separation being about 40 m. Moving into Chap. 4 next, the density and separation of the gas and dust grains will be found to respectively increase and decrease as we scrutinize, still further, the structure of molecular clouds, molecular cloud cores, and the nurseries of active star formation.

REFERENCES

1. In a groundbreaking study of Cepheid variables located within the Small and Large Magellanic clouds, Henrietta Leavitt (Harvard College Observatory), in 1908, found that their pulsation period varied in a systematic manner with brightness. Slowly pulsating, long-period Cepheids were intrinsically more luminous than more rapidly varying Cepheids. The identification of this relationship between brightness and pulsation period, once calibrated, has turned out to be of great importance with respect to the determination of distances to other galaxies. The RR Lyrae stars are just one of the several classes of calibrated pulsation variables.

2. While some galaxies are indeed spiral in appearance, galaxy morphology can take-on many diverse forms. Indeed, Edwin Hubble introduced, in 1926, his famous tuning-fork diagram to describe the morphological characteristics of the various elliptical, spiral and irregular galaxies. Hubble's classification scheme is based purely on visual appearance and only hints at the great diversity in galaxy morphology.

3. The amount of visual extinction is measured in terms of the so-called color excess $E(B - V)$, where the B and V magnitudes are determined through filters which allow full transmission of light at wavelengths of $\lambda = 0.436$ and 0.545 microns respectively – these are designated the blue and the visual filters. The color excess term is calculated according to measurements and calibration, with $E(B-V) = (B - V)_{observed} - (B - V)_{intrinsic}$. Here the idea is to exploit the fact that interstellar extinction not only reduces the total flux received at the telescope, but it also preferentially reduces the flux at shorter wavelengths – such fluxes being sampled through the B-filter. The color excess is then derived according to the observed $(B - V)$ color and the calibrated, or intrinsic $(B - V)$ color (for the spectral type of star being observed) when there is no extinction. The color excess can be expressed in terms of the absorption at each of the B and V wavelengths, so that $E(B - V) = AB - AV$, and from this the total selective extinction is determined, with $R_V = A_V/E(B - V)$, and the observations indicate that R_V varies from between 2 and 6, with $R_V \approx 3$ being typical of the diffuse interstellar medium. In a dense molecular cloud the relative extinction is higher and $R_V \approx 5$. The normalized extinction curve at any wavelength λ (i.e., that shown in figure 3.14) is expressed according to the ratio $A_\lambda/A_V = 1 + R_V (A_B - A_V) E(\lambda - V)/A_V$.

4. Continued observations of the motion of S2 (also S0-2) since 1995 reveal an orbital period of 15.2 years, and a closest approach distance to the central black hole of 120 astronomical units. At a distance of 7940 parsecs from the Sun, the orbital semi-major axis is estimated to be 980 AU and the orbital eccentricity is 0.876. At closest approach to the central black hole, S2 is traveling at some 5000 km/s, over 100 times faster than the typical space velocity of stars within the solar neighborhood. It is estimated that S2 has a mass of some 15 times that of the Sun, and accordingly, if not disrupted prematurely by its proximity to the central black hole, it will undergo supernova disruption sometime within the next ten million years. The orbit of S2 reveals (via Kepler's 3rd law) that the central black hole has a mass some 4.1 million times greater than that of the Sun, and such an object will have a Schwarzschild (event horizon) radius of 17 solar radii. For all its vast mass, the black hole at the center of our galaxy would sit comfortably interior to the orbit of Mercury within our solar system.

4

The Hyperfine Split and Atomic Jitters

Those melodies – that voiceful infinite!
And yet, they call it – Silence!

—John Pringle Nichol, *The Architecture of the Heavens* (1850)

Star Noise

"War! What is it good for?" So goes the 1970 protest song by Edwin Starr. Sad as it is to say, however, the answer to this question is that war is good for business, and it is good for scientific innovation. In less dramatic tones, perhaps, one could alternately say that "necessity is the mother of invention," and this is where science advances and knowledge benefits (?), during times of strife. And so it was that radio astronomy saw its origins as a direct result of World War Two.

Certainly one could argue, with a relatively high degree of confidence, that radio astronomy (and even space-based astronomy) would have come about without WWII and/ or the Cold War of the 1960s and 1970s, but the fact is that these conflicts brought together national interests and skilled scientists to solve specific tactical problems and perceived threats. With peace (of a sort) established in the mid-1940s these same scientists could take their previously secret research into more esoteric and open directions. Sir Bernard Lovell (Manchester University), for example, has written extensively on the early development of radio astronomy in Britain during the 1950s and 1960s, with the fledgling astronomers initially using ex-military radar and radio systems to study meteor showers, solar phenomena, and signal scintillation. Lovell has also described the political good will (with respect to monetary issues) that befell the then newly constructed (and now iconic) MK 1 steerable radio telescope[1] at Jodrell Bank (Fig. 4.1) as a result of its ability to track and monitor the radio "pings" of Sputnik-1 (launched on October 4, 1957).

Certainly radio astronomy, that is, the identification of extraterrestrial radio static, had taken its fledgling steeps several decades before WWII, with commercial industry, that

---

[1] The MK 1 was renamed the Lovell Telescope in 1987.

© Springer International Publishing AG 2017
M. Beech, *The Pillars of Creation*, Springer Praxis Books, DOI 10.1007/978-3-319-48775-5_4

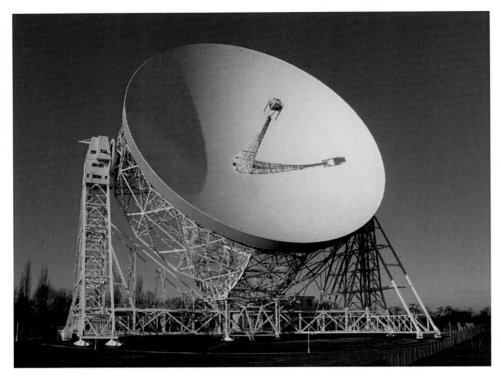

Fig. 4.1   The Lovell Telescope today. At 250 ft across it is still the world's largest fully steer-able radio telescope. Image courtesy of Jodrell Bank Centre for Astrophysics and the University of Manchester

other great driver of human history, looking to solve several specific problems associated with the transmission and reception of radio signals. The pioneer of these early investigations was Karl Jansky, a physicist in the employ of Bell Laboratories in New Jersey.

Jansky started work at Bell Labs in 1928, and by the early 1930s he had been charged with the task of identifying the reasons for a troublesome noise effect that was interfering with trans-Atlantic shortwave radio communications. To this end he constructed a large, steerable antenna tuned to a wavelength of 14.6 m (Fig. 4.2), and then he set about gathering data—lots of data—over the course of an entire year. By monitoring the static over many months, Jansky was able to rule out any terrestrial sources for the background hiss, and indeed, he was able to tie the "noise" to the Milky Way and specifically the galactic center in the direction of Sagittarius.

Jansky's now groundbreaking paper was read at the May 3, 1933, meeting of the Institute of Radio Engineers in New York. Remarkably, for what was essentially a science paper with the less than confident sounding title, "Electrical disturbances apparently of extraterrestrial origin," the work received wide coverage and publicity. Indeed, Jansky's paper was front page news on the May 5 issue of *The New York Times* newspaper, the article header reading: "New Radio Waves Traced to Center of the Milky Way: mysterious static, reported by K. G. Jansky, held to differ from cosmic rays." Likewise, *The Index-Journal* from South Carolina

Fig. 4.2 Karl Jansky and his "merry-go-round" antenna at Holmdel (It has often been suggested that Jansky would have received a Nobel Prize for his discovery of cosmic radiation had he but lived a few years longer (he died in 1950). Arno Penzias and Robert Wilson, who also worked for Bell Labs from the Holmdel site in New Jersey, received the Nobel Prize for Physics in 1978 for their 1964 discovery of the cosmic microwave background radiation.), New Jersey. Image courtesy of the NRAO

ran a front page article on Jansky's paper (sandwiched between an article concerning the eighth bombing that week of a Chicago restaurant, and an article about the death of a supposedly 197-year-old man in Ching-Yun, China), noting that, "there was no indication of any kind that the galactic waves represent some kind of interstellar signaling."

Indeed, it would seem that the world was ready for a story such as that being told by Jansky. In the depths of the Great Depression, towards the end of prohibition, and at a time when wonderment was very much in need, this new discovery, with its cosmic connection, appears to have acted as a social tonic. The story ran across the news media of the United States, and special events were soon being organized to hear the new cosmic hiss. Indeed, *The Brooklyn Daily Eagle*, New York, newspaper on May 16, 1933, reported that, "Radio waves [were] heard from remote space – sounding like steam escaping from a radiator." The article continued: "Listeners of station WJZ last night heard radio waves coming from a point so remote in space that it requires between 30,000 and 40,000 light years for them to reach the earth." *The Winnipeg Tribune* ran a similar such story concerning the broadcast of cosmic waves for October 19, 1933, describing it as making "a sizzling sound – like bacon in a frying pan."

The Hopewell Herald, New Jersey, ran yet another article concerning a broadcast in its November 15, 1933, issue, announcing that the Museum of Natural History would be broadcasting the cosmic signals for all to experience—"hear sound waves 40,000 years old" the trailer ran.

Not only did Jansky's galactic radio waves catch the attention of the general public, they also resulted in some remarkable speculations (wishful thinking) concerning the utility of the discovery. *The Emporia Daily Gazette*, Kansas, announced in its November 28, 1933, issue that, "a new dream of obtaining power from the stars of the milky way is opened to scientists in the form of radio waves streaming endlessly across space…. Karl G. Jansky listens to the signals made by the continual bombardment of the earth by star radio waves – straight from the Power House." The source of this particular speculation is not revealed, but it presumably harps back to the resonant inductive coupling experiments carried out by Nikola Tesla in the late 1890s.

The public world was abuzz with Jansky's discovery of radio waves, distant noise emanating from the galactic center, and astronomers were as equally excited to speculate about its origins, but ultimately the story failed. Perhaps it was the hard grind of the Great Depression or the gradual tailspin towards the political turmoil of WW II, but for whatever reasons, radio astronomy had a faltering start, the topic not seeing any real growth until the 1950s. The resurgence was driven by the establishment of new radio telescope observatories at many locations around the world—at Cambridge University and Jodrell Bank, under the guidance of Martin Ryle and Bernard Lovell, respectively; at Ohio State University, under the guidance of John Kraus; and at observatories in Australia, the Netherlands, and Canada. With the founding of these new observatories the rate of discovery and the development of new associations and insights increased dramatically. Literally a new and large universe came into focus.

During the lull between Jansky's initial study and the establishment of dedicated radio telescope observatories after WW II, there was one flickering beacon that kept the cosmic static torch flame alive. The carrier of the flame was American electrical engineer Grote Reber. Indeed, Reber was directly inspired by Jansky's discovery of cosmic static and at his own expense constructed a parabolic antenna (Fig. 4.3) and sensitive receiver system to study the heavens. Unlike Jansky, however, Reber set out to study cosmic noise over a wide range of wavelengths. His first receiver system was designed to work at a wavelength of 9 cm, but he failed to detect any noticeable signal. This non-detection was an immediately interesting result, and worked counter to expectation.

Given, as was then generally assumed, that the cosmic noise energy flux would follow that displayed by a blackbody radiator or thermal distribution, Reber had expected a much stronger signal strength at 9 cm than that recorded by Jansky at 14.6 m [1]. Reber reworked his receiver system, but again found no cosmic signal at a wavelength of 33 cm, nor at 62 cm. Finally, at a wavelength of 187 cm, Reber detected the distinctive all-sky hiss of cosmic noise. These pioneering observations revealed that the intensity of the cosmic noise signal increased in the exact opposite sense to that expected from a blackbody radiator, and this clearly indicated some new form of emission process must be at work.

Reber's early results were published in *The Astrophysical Journal* first in 1940 and then in 1944. These papers clearly established that the emission must be due to a non-thermal process, but they also provided the very first maps of the radio sky. In his 1944 paper,

Fig. 4.3 The parabolic radio telescope antennae constructed, in 1937, by Grote Reber in his backyard in Wheaton, Illinois. The sheet metal dish was 31.4 ft in diameter and steerable only in elevation. Image courtesy of NRAO

simply entitled "Cosmic Static," Reber published the first ever contour map of radio sky emission strength made at a wavelength of 190 cm. The emission contours clearly mapped out the circle of the Milky Way, and revealed strong and concentrated regions of emission towards the galactic center (in the direction of Sagittarius), and in the constellations of Cygnus and Cassiopeia. For these latter two sources there were no obvious optical counterparts, further mystifying the radio emission-generating mechanism. On the other hand, in his 1944 paper Reber presented the first direct evidence for radio emission from the Sun, but the signal was still very weak. These observations combined, as noted by Reber, indicated that emission from stars alone could be "discounted when explaining the origin of cosmic static."

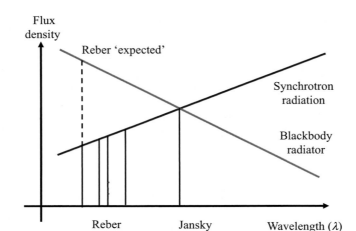

Fig. 4.4 Schematic variation (logarithmic scale) of energy flux density versus wavelength for synchrotron (non-thermal) radiation and a blackbody (thermal) radiator. The *vertical lines* indicate the wavelengths at which Jansky and Reber made their pioneering observations

If stars in general were not responsible for the generation of cosmic static, then what mechanism was actually at play? This was the question to answer, and as with many other episodes in human history when events are changing rapidly, and when new publications are appearing almost daily, which party did what and when has proved difficult for historians to unravel. In general, however, Russian physicist Vitaly Ginzberg[2] is usually credited with working out the detailed theory of the emission process behind cosmic static, in a series of research papers published during the 1950s. Indeed, cosmic static is the result of synchrotron radiation (recall Fig. 2.18). This form of radiation was first identified in the synchrotron (particle) accelerator laboratory constructed by General Electric in 1947. It is produced when a charged particle moves, in a helical fashion, along and around a magnetic field line. As described by the classical theory of electrodynamics, whenever a charged particle is accelerated it must emit electromagnetic radiation, and as the particle in the synchrotron case is moving along a spiral path, it is being continually accelerated and accordingly must continuously radiate electromagnetic radiation as it goes. Ginzberg (and independently by many others) showed that the power radiated by a charged particle, most often an electron, will vary according to the particles energy squared (as well as its mass and the magnetic field strength). Figure 4.4 shows the variation of energy flux density[3] versus wavelength for synchrotron radiation contrasted against that of a blackbody radiator.

---

[2] Ginzberg is perhaps better known for his (2003) Nobel Prize-winning work related to superconductivity and the remarkable physical properties of superfluids.

[3] The strength of a radio source is measured according to its energy flux density, which has units of Joules per second per meter squared per unit frequency. The standard unit in radio astronomy is, in fact, the Jansky, where $1~\text{Jy} = 10^{-26}~\text{J/s/m}^2/\text{Hz}$).

As the 1950s progressed radio astronomers began to finally pin down the precise locations of the stronger radio sources, and optical counterparts served as identifications. The strong radio sources identified by Reber in Cassiopeia and Cygnus were among the first objects to be sought out and identified by Walter Baade and Rudolf Minkowski, in 1954, using the newly constructed 200-inch Mount Palomar telescope (recall Fig. 2.15). Cassiopeia A, its location clearly identified by Martin Ryle and Francis Smith, turned out to be a supernova remnant. Located some 11,000 light years away, Cas A offers up one of those small but interesting mysteries of astronomy in that the supernova should have been visible from the Earth about 300 years ago, and yet no convincing accounts of any "new star" at that time have ever been found.

This result with respect to Cas A was consistent with the earlier 1949 identification by John Bolton, Gordon Stanley, and Bruce Slee, working from the Radiophysics Laboratory in Sydney, Australia, that the strong radio source identified as Taurus A was associated with the Crab Nebula. Cygnus A is one of the strongest radio sources in the entire sky, and it is an extragalactic object. Indeed, the emission is associated with, at optical wavelengths, an elliptical galaxy situated some 600 million light years away. Although Baade suggested that the host object was actually two galaxies in the process of collision, it is now generally thought to be a single galaxy with an accreting supermassive black hole situated at its core [2].

At radio wavelengths Cygnus A reveals a massive twin-lobbed emission structure (Fig. 4.5). The twin radio lobes stretch over a region of space some 200,000 pc across and are connected to the central elliptical galaxy by two oppositely directed jets. These jets are

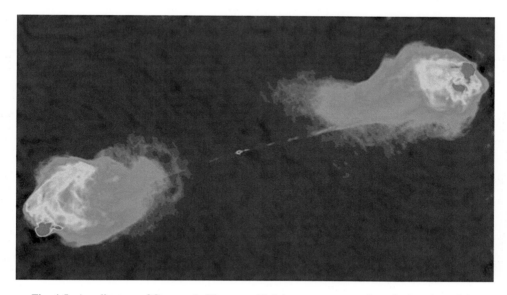

Fig. 4.5 A radio map of Cygnus A. The central bright spot corresponds to the location of the supermassive black hole located at the center of the host elliptical galaxy. The jets and large synchrotron emission lobes are clearly seen in the image, and two hot spots are produced where the jets ram into the intergalactic medium. Regions with the highest radio emission are colored *red*, while lowest intensity regions are shown in *blue*. Image courtesy of NRAO/AUI

produced through the accretion of material (disrupted stars and interstellar gas) onto a central black hole. The accretion process results in the production of high energy electrons that are then constrained to move along magnetic field lines that have been tightly wound up along the spin axis of the black hole. When first identified by Baade and Minkowski, the Doppler shift displacement of the optical spectrum lines of Cygnus A identified it as the most distant object then known. The Andromeda Galaxy (M31), at 2.5 million light years (778-kpc) distance, was first identified as a radio source in 1951.

Before moving on to discuss the pivotal role that the new and burgeoning developments in radio astronomy played in mapping out the interstellar medium and in delineating the galaxy's spiral arms, a few words should be said about cosmic rays. In the *New York Times* front page headline announcing Jansky's discovery of cosmic static it was stated that the new radio radiation was different to that observed from cosmic rays. What, therefore, are cosmic rays?

A quick look back at Fig. 2.16 indicates that cosmic ray astronomy began in the early decades of the twentieth century. Discovered and investigated by Victor Hess, cosmic rays, so named by physicist Robert Millikan, are not actually rays, as in the form of radiation, but high-energy particles. Cosmic rays travel at relativistic speeds, have an origin outside of the Solar System, and are predominantly hydrogen atom nuclei—that is, protons. A small percentage of cosmic rays are composed of electrons and atomic nuclei heavier than hydrogen. With respect to accelerating the particles to such high velocities, several mechanisms have been identified. The primary mechanism, however, is thought to be supernovae—an idea first suggested by the Baade and Zwicky duo in 1934.

Indeed, the Crab Nebula was recognized as an active source of cosmic rays by Yataro Sekido and co-workers in 1951. Cosmic rays intercept Earth's atmosphere all the time, although there are many, many fewer high energy cosmic rays than low energy ones. The higher energy cosmic rays can produce cosmic ray showers if they collide with an atom in Earth's atmosphere. Such collisions result in a great burst of secondary elementary atomic particles and their decay products. On the Earth's surface the consequences of such a collision can be detected through the occurrence of a brief, bright flash of Cherenkov radiation[4] either in the upper atmosphere or within ground-located water tanks.

Cosmic rays pervade the entire disk of the Milky Way Galaxy, and they can traverse vast distances, entrained by the galaxy's network of magnetic field lines (recall Fig. 3.16) from their site of origin. The lower energy, but much more abundant, cosmic rays play an important role in heating the interstellar medium. Such cosmic rays can penetrate deep into even the highest density clouds of interstellar gas, and through ionization and excitation interactions they can transfer energy to the surrounding medium. The heating takes place when a cosmic ray CR ionizes a hydrogen atom H, generating a free electron e^-, with the process proceeding as: $H + CR \rightarrow CR' + H^+ + e^-$, where CR' is the slightly less energetic cosmic ray that leaves the scene of the ionization event (the energy loss will relate to the electron ionization energy). It is the kinetic energy of the ejected electron that is eventually thermalized and converted into heat energy within the surrounding gas cloud.

---

[4] This form of radiation is produced when a particle moves through a medium at a speed greater than that at which light moves through it. While the speed of light in a vacuum c is a universal constant and an absolute limit, the speed of light through water, for example, is about ¾ c—accordingly it is entirely possible for a particle to have a speed greater than that with which light propagates within water.

Dialing Radio HI

Jan Oort was one of the most remarkable scientists of the twentieth century, and his legacy of ideas and astronomical discoveries is legendary. Oort spent most of his long research life at Leiden Observatory in the Netherlands, starting there as assistant director in 1934. His interests were broad and far reaching. He performed fundamental research relating to galactic structure, and identified the problem of missing mass (this now being called dark matter) and its association with galactic rotation in the early 1930s; in the 1950s he introduced the concept of the Oort Cloud, this being a massive spherical halo of some 10^{12} to 10^{13} cometary nuclei distributed in the region between 10^3 and 10^5 AU from the Sun. In the ecliptic plane the Oort Cloud extends inward to join up with the Kuiper Belt and the scattered disk of icy planetesimals and plutoids. The cometary nuclei in the Oort Cloud are only loosely bound by gravity to the Sun, and indeed, the outer boundary of the cloud effectively determines the limit of the Solar System.

Oort read Grote Reber's first 1940 *Astrophysical Journal* paper on cosmic static with great interest, and immediately realized that radio astronomy could afford astronomers a highly valuable research tool if emission lines, at radio wavelengths, could be identified. Oort suggested the radio emission line problem to a then young graduate student at the University of Utrecht, Hendrick van de Hulst, and in 1944 van de Hulst, at a colloquium held at Leiden, discussed the existence of a 21-cm wavelength emission line connected to neutral hydrogen.

The results of van de Hulst's talk and his theoretical analysis appeared in a three-page article entitled "Radiogolven uit het wereldruim" ("The Origin of Radio Waves from Space"), in the journal *Nederlands Tijdschrift voor Natuurkunde* (NTvN) in 1945. The paper was short in length but its content revolutionary, and it was destined to put post-WWII radio astronomy on the astronomical map.

Oort was made director of the Leiden Observatory in 1945, and with the closure of WWII hostilities he immediately set about establishing a strong group of radio astronomers in Holland. Radio telescopes were constructed at Kootwijk, Dwingeloo, and Westerbork, and the hunt was on for the 21-cm hydrogen line. Success followed in 1951, with two research papers (published alongside each other) appearing in the September 1 issue of the journal *Nature*. The first article was by Harold Ewen and Edward Purcell of Harvard University, while the second article was by Christiaan (Lex) Muller and Jan Oort. The article by Ewen and Purcell was one page long and simply reported the detection of the 21-cm line, with the first successful result being dated at March 25, 1951. Using the radio telescope at the Kootwijk Observatory, Muller and Oort detected the hydrogen emission line on May 11, 1951.

It was the cruel bite of misfortune that denied Muller and Oort from making the first accredited detection of the 21-cm emission line, since a fire destroyed their original apparatus. In spite of this early setback, however, the paper by Muller and Oort, while still only one-and-a-half pages long, was an altogether more substantial contribution than that by Ewen and Purcell. Indeed, combining Doppler shift velocity measurements with positional information Muller and Oort were able to infer properties of galactic rotation—a topic that they would go on to greatly develop over ensuing years.

The 21-cm hydrogen line comes about through electron transitions between the two hyperfine levels of the hydrogen ground state. The two hyperfine levels relate to a quantum mechanical property known as spin—a property that both the proton and the electron have in common. However, it transpires (and this is what van de Hulst first worked out) that when the spin of the electron and proton are anti-parallel the hydrogen atom is in a slightly lower energy state than when the two spins are parallel. A 21-cm emission line photon is emitted when the electron undergoes a spin-flip to put the hydrogen atom into its lowest possible energy state.

The energy associated with the hyperfine splitting of the hydrogen ground state (recall Fig. 2.8) is just $\Delta E = 5.9 \times 10^{-6}$ eV. The timescale for the spin-flip to occur in a given hydrogen atom can be further determined by quantum theory, and it is very long, being on average 3.5×10^{14} s (some 11 million years). One might at first think that given such a long timescale the spin-flip transition would only be seen very rarely, but since there are so many hydrogen atoms in the interstellar medium (about 10^{67}, in fact) the spin-flip transition, in multitudes of hydrogen atoms, is always taking place, and the end result is that the emission line is readily observable.

Given an energy ΔE for the electron spin-flip transition, the emission photon will have a frequency f corresponding to $\Delta E = h f$, where h is Planck's constant. Accordingly the hyperfine spin-flip transition is associated with a photon of frequency $f = 1.420405751786$ GHz, and this gives the wavelength of the transition photon as $\lambda = 21.10611405413$ cm. These numbers for the frequency and wavelength are worth seeing in their full decimal glory, since as noted by Carl Sagan and Frank Drake in the early 1970s they are fundamental numbers that are in principle knowable to every intelligent civilization in the entire universe. Indeed, Sagan and Drake used the universal applicability of the hyperfine transition constants to define base units for time (corresponding to $1/f = 7.0403 \times 10^{-10}$ s) and distance (corresponding to $\lambda_{21\,cm}$). These same base units were encoded into the plaques (Fig. 4.6) placed on the *Pioneer 10* and *Pioneer 11* spacecraft launched in 1972 and 1973, respectively.[5]

The importance of the 21-cm hydrogen line was further emphasized by Giuseppe Cocconi and Philip Morrison in their groundbreaking 1959 paper, "Searching for Interstellar Communications." Realizing that radio communication is probably the easiest way that any extraterrestrial civilization could identify itself to the rest of the galaxy, Cocconi and Morrison suggested that the hydrogen line offers a natural wavelength upon which to broadcast [3]. In terms of the essentially infinite range of wavelengths that one could broadcast a signal, the hydrogen line is the analog of a watering hole in the desert. Everyone will eventually look there [4].

Since the paper by Cocconi and Morrison appeared other researchers have suggested that additional interstellar communication wavelengths might be based upon the irrational,[6] and therefore highly special, numbers such as π, $2^{\frac{1}{2}}$, φ, and e. Perhaps the most important message to be found in Cocconi and Morrison, however, lies within the very last sentence of their paper, which reads: "[T]he probability of success is difficult to estimate, but if we never search the chance of success is zero."

---

[5] At the time of writing *Pioneer 10* is located 116 AU from the Sun, while *Pioneer 11* is 95 AU away. They have a very, very long way to go yet before they truly begin to sail into interstellar space.

[6] An irrational number is one that cannot be expressed through the division of two integer numbers.

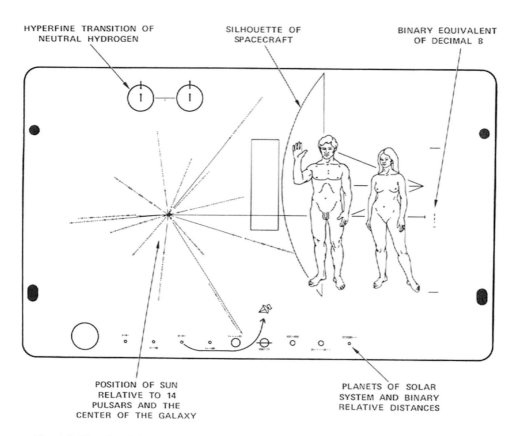

HYPERFINE TRANSITION OF
NEUTRAL HYDROGEN

SILHOUETTE OF
SPACECRAFT

BINARY EQUIVALENT
OF DECIMAL 8

POSITION OF SUN
RELATIVE TO 14
PULSARS AND THE
CENTER OF THE GALAXY

PLANETS OF SOLAR
SYSTEM AND BINARY
RELATIVE DISTANCES

Fig. 4.6 The pictorial message plaque carried on the *Pioneer 10* and *11* spacecraft. The message was developed by Carl Sagan and Frank Drake on the basis that the spacecraft might one day be intercepted by an extraterrestrial intelligence. At the *top left* of the plaque is a schematic image of the hyperfine transition for neutral hydrogen, and it is indicated in binary notation that this transition sets the time and distance units. The Sun position is intercepted by 15 radial lines. The *long horizontal line* indicates the distance of the Sun from the galactic center (this sets the pulsar map scale with $8 \equiv 8 \text{ kpc} \approx 10^{21} \lambda_{21\,\text{cm}}$ units), while the other 14 lines indicate the location, distance to (by the scale length of the line), and the periods of 14 pulsars. In principle this information would allow a sufficiently advanced extraterrestrial intelligence to locate the Sun within the galaxy. At the *bottom* of the plaque is a schematic representation of the Solar System. The male and female figures are shown to scale with respect to the spacecraft. Image courtesy of NASA

With the 21-cm emission line detected, the next task for the radio astronomers was to map the neutral hydrogen distribution of the Milky Way Galaxy. The Dutch group was at the forefront of this work, and in 1945 van de Hulst, Muller, and Oort produced the first neutral hydrogen map revealing, for the first time, the spiral arm structure within the disk of the galaxy. Radio astronomers in Australia additionally surveyed the sky at 21 cm, and in 1958 a combined Leiden-Sydney map of neutral hydrogen in the galactic disk was

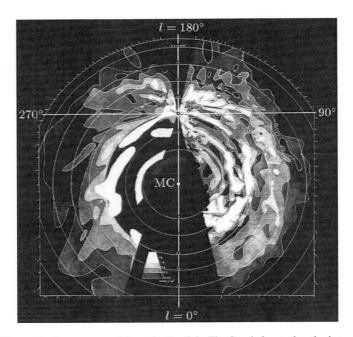

Fig. 4.7 Neutral hydrogen map of the galactic disk. The Sun is located at the intercept of the galactic longitude coordinate lines; MC indicates the galactic center. A grayscale has been used, with highest emission strength (gas density) being indicated by those regions shown in *white*. The spiral arm-like distribution of the neutral hydrogen is evident in the data but requires a certain amount of faith to see it with the eye. Compare and contrast this diagram with Fig. 2.14

published by Oort, Frank Kerr, and Gart Westerhout in the January 10 issue of the *Monthly Notices of the Royal Astronomical Society* (Fig. 4.7). The production of this map was a non-trivial exercise and entailed the measurement of the signal intensity along numerous lines of sight in galactic longitude. The signal strength gives an indication as to how much neutral hydrogen there is in the line of sight, and this combined with a galactic rotation model and the Doppler-shifted profiles of individual clumps enabled a direction, distance and hydrogen density map to be built up.

Oort et al. write in their 1958 review paper that, "[T]he distribution of the hydrogen evidently shows great irregularities. Nevertheless, several arms can be followed over considerable lengths." The Sun is located some 8000 pc (26,000 light years) from the galactic center (recall Shapley's result from Chap. 3) and is located on the outer edge of the Orion-Cygnus spiral arm spur, which in turn is sandwiched between the Sagittarius-Carina and Perseus spiral arms (recall Fig. 2.14).

Now that the neutral hydrogen content of the galaxy was observable (with radio telescopes), estimates of its specific mass fraction could be made. The modern-day numbers indicate that 70 % of the mass of the gas component of the interstellar medium is in the form of hydrogen (28 % of the mass is made up of helium). By number, rather than mass, hydrogen atoms make up about 91 % of all the atoms in the interstellar medium, with helium atoms contributing nearly 9 % to the rest.

By mass about 1.5 % of the interstellar medium is made up from atoms other than hydrogen and helium. These so-called heavy elements are synthesized in stars and supernova explosions (see Chap. 5), and it is through these latter atoms that complex molecules can be built up. Indeed, the existence of at least simple molecules in the interstellar medium was evident as early as the 1930s, but it was through the advent of new radio astronomy techniques, introduced during the late 1960s and early 1970s, that astronomers began to fully appreciate just how important the molecular cloud component of the interstellar medium was, and indeed is, with respect to the process of star formation.

The Molecular Giants Emerge

Although simple molecules had been identified in stellar atmospheres very early on in the history of spectroscopy, it was somewhat of a surprise to find molecules in the interstellar medium. Indeed, it was generally thought in the early 1900s that any rare molecules that might chance to form in the interstellar medium would soon be dissociated by UV radiation. What had not been appreciated at that time was the important shielding role played by interstellar dust. Nonetheless, in 1937 Pol Swings and Leon Rosenfeld somewhat tentatively identified an interstellar absorption line due to CH (methylidyne) superimposed upon a series of stellar spectra obtained by Theodore Dunham. Shortly thereafter, in 1940, Andrew Mckeller detected interstellar CN (cyano radical) lines superimposed upon a series of stellar spectra, with Alex Douglas and Gerhard Herzberg adding interstellar CH^+ (methylidyne cation) to the list in 1941.

The first detection of an interstellar molecule with a radio telescope was made by Sander Weinreb and coworkers from MIT in 1963, who found the emission line of OH (hydroxyl radical) in the supernova remnant Cassiopeia A. Other interstellar OH detections in HII regions soon followed, and in 1965 radio astronomers at Berkeley, led by Harold Weaver, not only identified OH emission lines in the direction of the Orion Nebula, and in the direction of an HII region called W3 in the constellation of Cassiopeia, but also a series of sharp and strong emission lines that they could not identify. These lines they attributed to an unknown element they called "mysterium." It was soon realized, however, that the mysterium lines were actually due to OH, but OH in a highly perturbed state. Indeed, the mysterium lines were due to an OH maser [5].

Through the 1960s and 1970s more and more interstellar molecules were found by radio astronomers (Fig. 4.8). Lewis Snyder and co-workers found interstellar formaldehyde (H_2CO) in 1969. Importantly for later studies Arno Penzias, Keith Jefferts, and Robert Wilson observed interstellar CO in the Orion Nebula in 1970, and in the same year George Carruthers identified lines due to interstellar molecular hydrogen H_2. As of today well over 200 interstellar molecules have been identified.

In the context of astrobiology and the origin of life debate (see Chap. 6) the discovery of CH_2OOCHO (glycolaldehyde) by Jan Hollis and coworkers in 2000 marks an important milestone. The detection was made towards the galactic center in a region called Sgr B2(N) (see below), and the importance of glycolaldehyde with respect to the origin of life

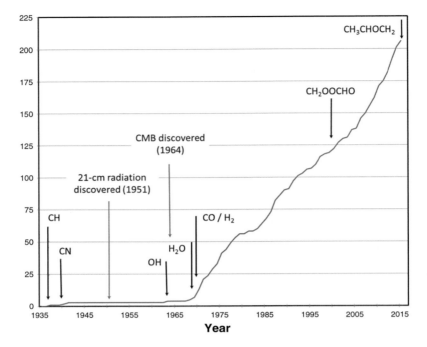

Fig. 4.8 The cumulative number of detected interstellar molecule species versus time. The detection time of various key molecules are indicated by *arrows*. The times at which the 21-cm hydrogen line and the cosmic microwave background (CMB) radiation were discovered are shown for reference. Data from www.astrochymist.org

debate is that it is the simplest possible member of the monosaccharide sugars. Laboratory experiments have shown that glycolaldehyde can be converted into amino acids (the building blocks of deoxyribonucleic acid—DNA), and it is generally thought that this molecule must have played an important role in early abiogenesis by aiding in the synthesis of ribose, the later carbohydrate being a fundamental component of ribonucleic acid (RNA). The importance of RNA is that it is generally thought to have mediated the replication pathway through which the first life on Earth evolved. The RNA would kick-start the process, as it were, before DNA would take over with respect to information storage and replication.

The discovery of another important complex molecule, propylene oxide (CH_3CHOCH_2), was announced by Brett McGuire (NRAO) and coworkers in the June 14, 2016, issue of *Science*. This molecule, like glycolaldehyde, was detected in Sgr B2 and has important implications for the origin of life debate since it has what are called chiral properties. Indeed, propylene oxide comes in a left- and right-handed version (this is the chiral characteristic), and this property of handedness has been exploited in many biological structures. Amino acids and sugar molecules, for example, have important chiral characteristics. Industrial biochemists have exploited the chiral properties of milk sugar in developing artificial sweeteners, with the artificial sweeteners using the left-handed version of the molecule. Since our bodies can only absorb the right-handed sugar molecule, the artificial sweetener can be used for its taste, but zero calories are actually ingested. Ibuprofen and warfarin are also important drugs with chiral properties.

The vitally important role played by interstellar molecules in the origin of life debate is the subject of Chap. 6. In terms of the most massive interstellar molecules to be discovered so far, though, the record is presently held by C_{60} and C_{70}, in which 60 and 70 carbon atoms, respectively, are involved in the construction of fullerene, or so-called buckyballs. C_{60} is a molecule made of 60 carbon atoms arranged to form a hollow spherical cage-like structure that was first synthesized in the laboratory in 1985. Called Buckminsterfullerene, after the American architect Buckminster Fuller, who perfected the design of geodetic domes in the 1950s, C_{60} was first detected in space by Jan Carmi (University of western Ontario, Canada) and coworkers in 2010. At infrared wavelengths, in fact, detected by the Spitzer Space Telescope, C_{60} and C_{70} (made of 70 carbon atoms arranged in the cage-like shape of a hollow ellipsoidal shell) were identified by Carmi's group in the planetary nebula Tc 1 (the first object in the catalog by Andrew Thackery, Radcliffe Observatory in South Africa). This particularly young planetary nebula surrounds an evolved carbon-rich star losing mass into surrounding space, and the emission lines of C_{60} and C_{70}, for presently unclear reasons, are particularly strong (Fig. 4.9).

The study published by Carmi et al. indicates that emissions from the fullerenes did not originate from free molecules in a gas phase but were linked to molecular carriers anchored to solid carbonaceous dust grains, the dust grains having previously formed in the

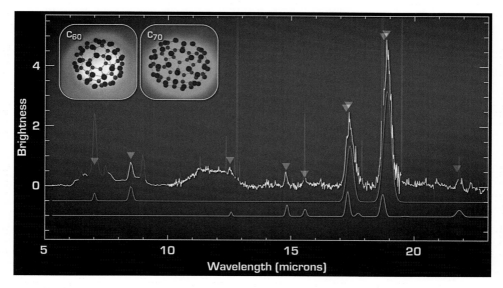

Fig. 4.9 Detection of emission lines due to C_{60} and C_{70} fullerene in the direction towards planetary nebula Tc 1 (=He2-274 = PK 345-8°1 = IC 1266) in the constellation of Ara. The emission line spectrum was recorded with the Spitzer Space Telescope. *Insets to the upper left* illustrate the atomic structure of the C_{60} and C_{70} fullerene. Image courtesy of NASA/JPL-Caltech/UWO/SETI Institute

carbon-rich outflow from the planetary nebula's central star. The process by which molecules can grow on the surface of interstellar grains will be discussed in more detail later.

Molecular emission and absorption line spectra are much more complex than those produced by single atoms. Not only do molecules undergo electron transitions, with the electrons moving between allowed orbits, but they can also produce rotation and vibration lines. As described before, the electronic transitions are readily observable in the optical and UV part of the electromagnetic spectrum. The rotation lines are produced when the collective motion of the atomic nuclei in the molecule changes from one quantum state to another, and these lines typically fall in the microwave to millimeter region of the electromagnetic spectrum. Vibrational lines are produced through the relative motion of the atomic nuclei, and these lines tend to fall in the infrared region of the electromagnetic spectrum.

Both the rotation and vibration modes of a molecule are governed by quantum mechanical selection rules. The energy level of a specific molecule in the Jth rotation state is written as $E_J = J(J+1)B$, where B is a constant (the so-called rotational constant) that varies from one molecular species to the next. For CO, it is $B = 1.93$ cm^{-1}; for H_2, it is $B = 60.80$ cm^{-1}. When the rotational state of a molecule changes from the J to $J-1$ level, a photon of energy $\Delta E = E_J - E_{J-1} = 2BJ$ will be emitted, so for the $J = 1 \rightarrow 0$ transition, the emission line will have an energy $2B$, for the $J = 2 \rightarrow 1$ transition the emission line will have energy $4B$, for $J = 3 \rightarrow 2$ the energy is $6B$, and so on.

Accordingly for CO, the first three rotational emission lines will appear at wavelengths of 2.60, 1.30, and 0.867 mm, respectively. One of the great benefits of having a whole series of rotational state emission lines to measure is that the temperature structure of a molecular cloud can be studied. If the temperature of the gas is high, then collisions between atoms and molecules will have plenty of energy to place the molecules into high rotation states with $J = 3$ or 4, for example. In cooler regions the collisional energy will only place molecules in relatively low rotational states with say, $J = 0$, 1 or 2. The more molecules there are in a specific rotation state, the stronger will the signal intensity be at its associated emission wavelength. Hence, by comparing (say) the emission line strength of the $J = 3 \rightarrow 2$ transition at 0.867 mm wavelength to the $J = 1 \rightarrow 0$ emission line at 2.6 mm wavelength, the temperature structure of a molecular cloud can be analyzed.

Hydrogen is by far the most abundant species of atom in the interstellar medium, and under the right conditions it can form the H_2 molecule. Accordingly, it would be expected that the most abundant molecule in the interstellar medium is likely going to be H_2. Showing this observationally, however, has proved difficult. The problem with H_2 is that it is a symmetrical molecule made up of two identical atoms, and this dictates that the selection rules for both the rotational and vibrational modes result in weak and hard to observe emission lines.

Although H_2 is in fact the most abundant molecule in the interstellar medium it is virtually invisible to direct observations. The existence of H_2 molecules can be inferred in the line of sight, however, via its production of absorption lines in the ultraviolet part of the spectrum. Such absorption lines were first observed in 1970 in the spectrum of the star Xi-Persei (a young, massive, and highly luminous O7.5 spectral-type blue supergiant star in the constellation of Perseus, with the Arabic name Menkib). It is this particular star that powers the HII region known as the California Nebula (NGC 1499).

The UV spectrum of Xi-Persei was recorded during an Aerobee-150 rocket flight and analyzed by George Carruthers (U. S. Naval Research Laboratory, Washington, D. C.), who concluded in his research paper published in the August 1970 issue of *The Astrophysical Journal* that, "nearly half of the total hydrogen in the line of sight to Per[seus] may be in molecular form." Subsequent observations have confirmed that H_2 is by far the most abundant molecule in interstellar space, but most contemporary research infers the actual abundance of H_2 via tracer molecules that are more straightforward in terms of observing with radio telescopes. Among the most important tracer molecules for H_2 is the carbon monoxide molecule, CO, which produces (as seen above) strong rotation lines at wavelengths of 2.6 and 1.3 mm.

Interstellar CO lines were first detected in radio telescope observations of the Orion Nebula (M42) by Arno Penzias, Keith Jefferts, and Robert Wilson in 1970. This research triumvirate also identified CO lines in the star-forming region M17 (the Omega Nebula) and seven other locations situated close to the galactic center, including those regions identified as Sgr A (which enshrouds the supermassive black hole at the galactic center, identified as the strong radio source A*) and the giant molecular cloud, identified as Sgr B2, located some 390 light years from the galactic center.

The Orion Nebula has featured throughout this text as one of the most iconic astronomical objects in the universe, and it was an obvious early target for radio astronomers. The nebula was first detected at a wavelength of 9.4 cm by Frederck Haddock (U. S. Naval Research Laboratory, Washington, D.C.) and coworkers in November 1954, and it was mapped out at a wavelength of 10 cm by T. K. Menon (NRAO) in 1962. Figure 4.10 shows a 20-cm wavelength radio map of the Orion Nebula as published by Marcello Felli (Osservatorio Astrofisico di Arcetri, Italy) et al. in 1993. The latter map brings out the distinct linear feature of the Orion bar, where the stellar wind and ionization front produced by the Trapezium stars (recall Fig. 1.9) pushes into the surrounding molecular cloud.

Penzias and coworkers made the first detection of CO in emission across the Orion Nebula at a wavelength of 2.6 mm in 1970, and Fig. 4.11 shows a more recent CO map produced by Ronald Maddalena and coworkers in 1996. The 30×40 degree map by Maddalena et al. reveals that the Orion molecular cloud is a much more extensive and complicated structure than that of the HII emission region associated with the Trapezium stars. The Orion molecular cloud has several distinct components that have been producing new stars for at least the past several millions of years. From images such as CO map in Fig. 4.11 we see that the star-forming H II regions (i.e., M42) are really just the more obvious visual components of much more massive and spatially extensive molecular cloud structures.

There are many sub-components to the Orion molecular cloud, but the so-called Orion A and Orion B clouds are particularly distinctive. The Orion Nebula is contained within the Orion A cloud. Combining infrared and CO radio telescope observations, Frank Ripple (University of Massachusetts) and coworkers estimated in a 2013 research publication that the amount of atomic and molecular mass contained within Orion A and Orion B is 63,100 and 46,300 solar masses, respectively.

In addition to the CO observations, the Orion molecular cloud complex has been mapped out at far infrared wavelengths, and these latter observations reveal the presence of multiple molecular emission lines. Indeed, in 2010 observations gathered by the

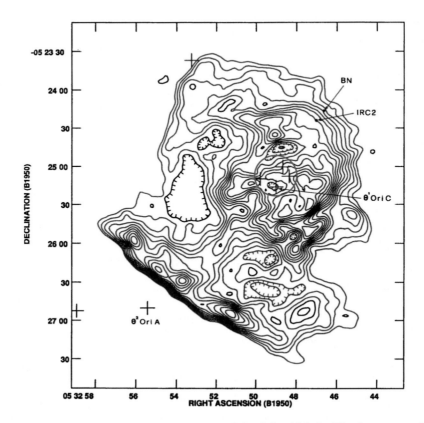

Fig. 4.10 A 20-cm wavelength emission map of the Orion Nebula. The *four center-right crosses* indicate the position of the Trapezium stars (*red circle*), with theta-1 Orionis C being labeled. IRC2 and BN correspond to prominent infrared emission regions. The strong linear feature to the *lower left* in the image, above theta-2 Orionis A, is known as the Orion bar, and is a region of high gas compression. Image adapted from Filli et al. 1993

Herschel Space Observatory revealed multiple emission lines due to many organic chemicals deemed essential to the evolution of life (Fig. 4.12). For example, the Herschel observations reveal distinct emission lines due to water, carbon monoxide, formaldehyde, methanol, dimethyl ether, hydrogen cyanide, sulfur oxide, and sulfur dioxide. How and where the molecules found within molecular clouds form will be discussed later in this chapter.

Molecular emission line surveys, primarily in CO lines, have revealed that molecules within the interstellar medium are mostly clumped together into large clouds, spread over several to several hundred light years, with masses that can vary from a few to as much as many millions of solar masses. Such giant molecular clouds (GMCs) have number densities that vary from 10^{11} to 10^{12} particles (atoms/molecules/dust grains) per cubic meter.

Invariably GMCs have a complex filamentary structure and show dense clumps—called molecular clumps or cores—with densities up to a thousand times greater than that of the surrounding GMC. For molecules to survive against UV dissociation they need some form

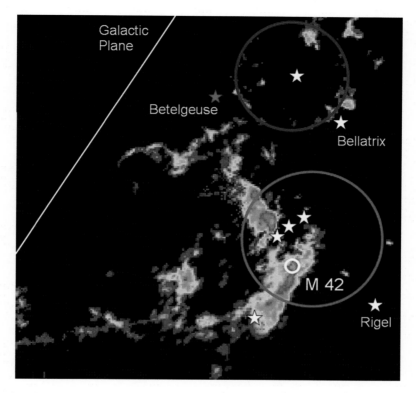

Fig. 4.11 CO map of the Orion Nebula (M42). The CO emission levels (*red* being highest emission) reveal that the Orion molecular cloud is very much more extensive than the Orion Nebula itself. The positions of the principal stars in the constellation of Orion are highlighted for reference. Barnard's Loop (recall Fig. 2.10) is shown in *red*, and the *blue circle* indicates the location of the λ Orionis Nebula. The Orion A cloud is towards the *lower right* in the image, and the Orion B cloud is towards the *center*. CO background map courtesy of NRAO/Maddalena

of shielding, and this is provided for by interstellar dust grains. The dust grains not only provide shielding for the molecules, they also provide the all-important quiescent surfaces where the molecules can actually form. The dust literally provides the factory floor where captured atoms can accumulate, migrate, and build up complex structures.

As indicated earlier, CO is used as a tracer molecule for molecular hydrogen, and indeed it is through collisions with H_2 molecules that CO emission lines are produced. The higher the density of H_2 in a particular region, the higher the so-called velocity integrated CO emission intensity W_{CO}. Usefully, the column density of molecular hydrogen $N(H_2)$ along a specific line of sight can be expressed as the approximate constant ratio: $N(H_2)/W_{CO} = 2 \times 10^{20}$ molecules per square centimeter per Kelvin per km/s. The somewhat odd-looking units for the $N(H_2)/W_{CO}$ ratio relate to the manner in which the integrated intensity W_{CO} is determined. This quantity being expressed as the brightness temperature (in Kelvin) integrated over the radial velocity (km/s) variation encompassed by the emission line profile.

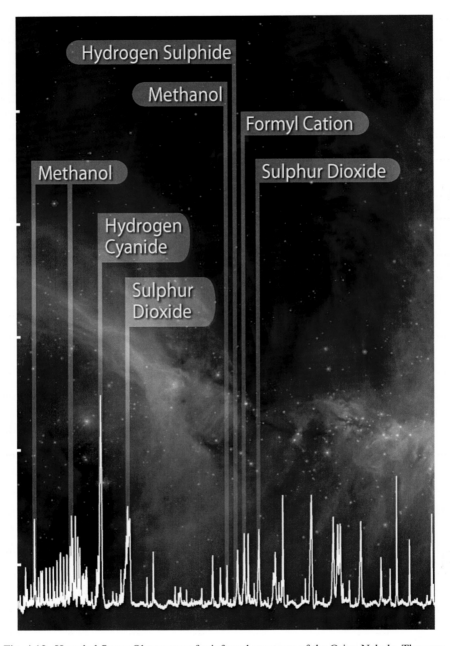

Fig. 4.12 Herschel Space Observatory far infrared spectrum of the Orion Nebula. The spectrum indicates that M42 is a prolific molecule-making factory. Image courtesy of ESA, HEXOS, and the HIFI Consortium

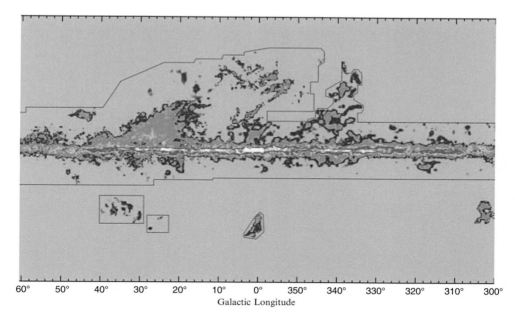

60° 50° 40° 30° 20° 10° 0° 350° 340° 330° 320° 310° 300°

Galactic Longitude

Fig. 4.13 Segment of the velocity-integrated CO emission map along the plain of the Milky Way. The image center indicates the location of the galactic center. Image courtesy of the CfA Millimeter-Wave Group

The first survey maps at CO wavelengths were published in the mid- to late-1970s, and they began to reveal a detailed picture of the large-scale distribution of molecular clouds within the galactic plane. One of the first comprehensive maps of the galactic plane to be published was that by Thomas Dame (Harvard-Smithsonian Center for Astrophysics) and coworkers in 1987. Figure 4.13 shows a more recent galactic plane survey map published by Dame et al. in 2001. Based upon 37 individual CO survey sets, produced over some 20 years of study, it is evident that molecular clouds are tightly concentrated towards the galactic plane, typically being no higher than 50 pc above or below the galactic mid-plane.

Additionally, when the velocity versus galactic longitude data is considered (see Fig. 4.14), it is apparent that most of the molecular clouds are located interior to the Sun's orbit and situated within a giant ring-like feature (the 5 kpc ring), with inner and outer radii between 3 and 7 kpc centered on the galactic core (Fig. 4.15). It turns out that most of the star formation within the galaxy takes place within the 5 kpc ring. Indeed, with an estimated mass of several billion times that of the Sun, the molecular clouds within the 5 kpc ring alone contain about 70 % of all the molecular gas interior to the Sun's orbit, and it affords the galaxy a great storehouse of material out of which future generations of stars will be born.

Interior to the 5 kpc ring is the central molecular zone (CMZ), a region of very complex structure, delineated by dense molecular clouds, massive star formation regions, supernova remnants, and high velocity gas streams (Fig. 4.16). The CMZ stretches outward to about 1000 light years from the galactic center, but also reveals additional complex

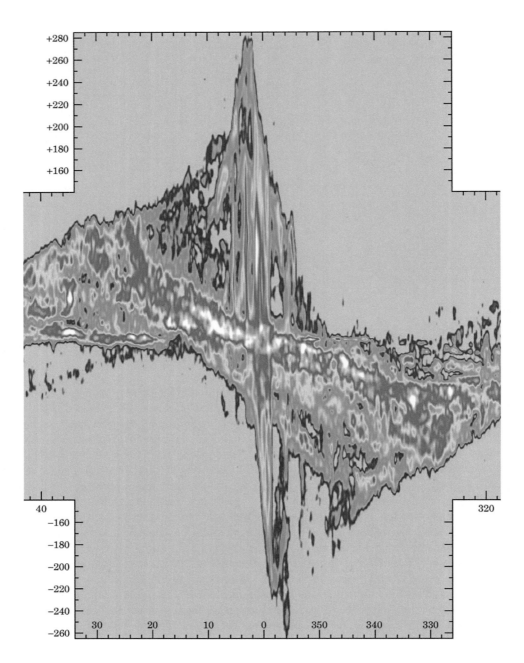

Fig. 4.14 Segment of the galactic longitude versus velocity map of CO emissions. The central slanting bar in the data corresponds to the 5 kpc ring material, while the central spike indicates an inner, rapidly rotating disk with a radius of some 500 pc (1500 light years). Image courtesy of the CfA Millimeter-Wave Group

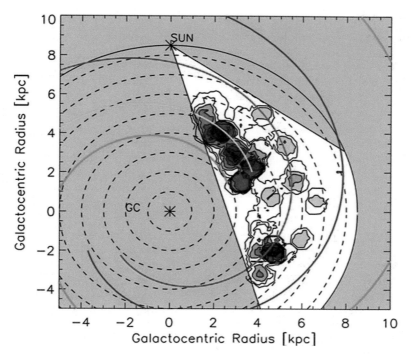

Fig. 4.15 Galactic ring survey detail. The image shows the surface mass density of molecular clouds situated between galactic longitudes of 18 and 52°. The ring-like distribution of molecular clouds between 3 and 6 kpc is evident in the survey data, although the molecular clouds additionally trace out regions delineated by the Scutum and Sagittarius spiral arms. The *green, yellow, red, and blue spiral arcs* represent the 3 kpc, the Scutum-Crux arm, the Sagittarius arm, and the Perseus arm, respectively. Image adapted from Julia Roman-Duval, Institute of Astrophysical Research at Boston University, et al., 2010 *ApJ.* **723**, 492

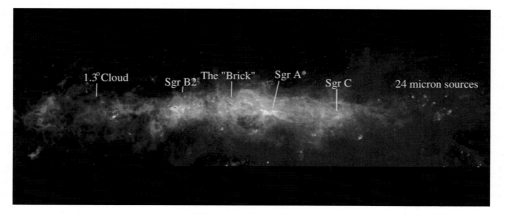

Fig. 4.16 The central molecular zone. This is a composite infrared and radio wavelength image, with dense gas being shown in *red*, warm and cold dust in *green and blue*. The central black hole is located in the region labeled Sgr A*. To the *far right* of the image can be seen a series of dust enshrouded young stellar clusters observed as prominent infrared point sources at a wavelength of 24 μm. Image courtesy of C. Battersby (Smithsonian Astrophysical Observatory, 2016

substructures. The innermost region, some 10 light years in diameter, for example, forms a doughnut-like structure around the central black hole, although for reasons that are still not clear the CMZ is not centered on Sgr A*.

A recent study by Jonathan Henshaw (Liverpool John Moores University) and coworkers draws attention to the observation that though the CMZ contains less than 10% of the Milky Way's molecular gas, it contains some 80% of the densest gas clouds with $n_{gas} > 10^{12}$ m^{-3}, representing a deep reservoir of some 40 million solar masses worth of molecular material, and yet the zone shows relatively low levels of active star formation. Henshaw and coworkers, writing in the April 11, 2016, issue of the *Monthly Notices of the Royal Astronomical Society*, suggest that star formation activity is reduced in the CMZ because of the highly turbulent motions exhibited by the molecular gas, and they additionally trace a number of new supersonic gas-flow streams across the CMZ.

Also identified in Fig. 4.16 are several distinctive giant molecular clouds: Sgr B2 is some 50 pc across and contains about 3 million solar masses of molecular gas, and stars are actively forming within its inner core. The Sgr C complex is composed of several giant molecular clouds and star formation regions. Mapping of the $l = 1.3°$ molecular cloud has recently been carried out by Takashi Oka et al. with the Atacama Submillimeter Telescope Experiment (ASTE), and these observations indicate that the region contains a whole host of expanding gas shells, presumably the result of supernova explosions, and this suggests that a massive stellar cluster must be embedded within its central core. The feature labeled as the Brick (less dramatically identified as GO.253 + 0.016) in Fig. 4.16 is a very dense molecular cloud; indeed, it is one of the highest density regions within the entire Milky Way Galaxy, and packs some 200,000 solar masses of molecular gas into a region that is just 8 pc across.

Remarkably, for all of its mass, the Brick shows no evidence of any embedded HII regions, and it appears to have produced hardly any stars at all. Jens Kauffmann (California Institute of Technology) and coworkers suggest in a 2013 publication in *The Astrophysical Journal* that this latter result implies that the Brick is a very young molecular cloud that has yet to dissipate its strong internal gas motions and begin the process of making stars.

Moving away from the galactic center, the distribution of molecular clouds within 1 kpc (3300 light years) of the Sun is shown in Fig. 4.17. Inside this disk, stretching some 50 pc above and below the galactic plane, are 21 clouds containing some 4×10^6 M$_\odot$ of molecular gas. Accordingly, in the same volume of space, the total (H$_2$ + HI) mass within 1 kpc of the Sun is about a quarter of that held within the stars. About half of the total molecular mass is contained within five compact clouds, identified as the Cygnus Rift, Cygnus OB7, Cepheus cloud, Orion B, and Monoceros OB1. Each of these clouds contains several hundred thousand solar masses of molecular gas.

The nearest molecular cloud to the Solar System at the present epoch is TMC-1, located in the direction of the constellation Taurus. This cloud is located some 140 pc away and contains some 30,000 M$_\odot$ of molecular gas. The most massive molecular cloud within 1 kpc of the Sun is that of the Cygnus Rift, which contains an estimated 1 million solar masses of molecular gas. The Cygnus Rift, often called the Northern Coalsack (in contradistinction to the Coalsack Nebula in the southern hemisphere) stretches from the constellation of Cygnus all the way through to Sagittarius, marking out a long strand of dust-rich molecular clouds that divides the arch of the Milky Way into two glowing halves. Indeed, the Cygnus Rift is both visible and conspicuous to the unaided human eye.

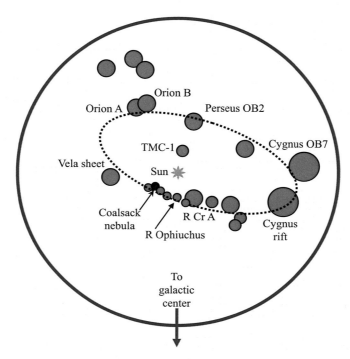

Fig. 4.17 The distribution of giant molecular cloud structures within 1 kpc of the Sun. The *dashed ellipse* indicates the approximate extent of Gould's Belt. The size of the molecular clouds is indicative of total mass. The three circle sizes used indicate masses of order 10^3, 10^5, and 10^6 $M_\odot$, respectively. Diagram modified from Dame et al. 1987

Looking closely at Fig. 4.17, the eye finds the strong hint of an elliptical ring-like distribution of molecular clouds and star-forming regions with an approximate radius of some 500 pc around the Sun. This hint is further enhanced, it turns out, if one additionally looks at the distribution of young, luminous, and massive stars within the solar neighborhood. Indeed, it was by looking at the latter distribution of bright stars that Benjamin Gould[7] identified the feature in 1879, noting specifically that a "great circle or zone of bright stars seems to gird the sky." Now known as Gould's Belt, the origin of this ring-like distribution is presently unclear—it may be just a coincidence, or it may represent a spiral arm feature, or indeed, it may be the result of a dark matter *bullet* passing through the galactic disk.

In general terms Gould's Belt is inclined by some 20° to the galactic plane, is about 1000 pc across and contains of order a million solar masses' worth of stars and molecular gas. The Sun is not at the center of the belt, but is offset by some 200 pc; it additionally appears that the belt is slowly expanding outward. An analysis of the constituent stars and star clusters situated around the belt indicate a formation age set some 30–50 million

---

[7] Gould is additionally remembered for being the founding editor of the *Astronomical Journal*.

years ago. A recent 2015 study of the distribution of young O and B spectral-type stars within 500 pc of the Sun, by Hervé Bouy (CSIC-INTA, Madrid) and João Alves (University of Vienna), appears to map out the Gould Belt structure, but the authors argue that the belt is really a projection effect produced by three elongated stream-like structures that, purely by chance, are arranged in what appears to be a ring-like feature. Other researchers have suggested that Gould's Belt is the result of an induced star formation event triggered by an external impact—either by a high velocity galactic-halo gas cloud or a dark matter clump. Kenji Bekki (University of New South Wales, Australia) first discussed this latter scenario in some detail in 2009, suggesting that the belt is the result of a collision of a $10^7 \, M_\odot$ dark matter clump and a $10^6 \, M_\odot$ giant molecular cloud some 30–35 million years ago.

Interestingly, since this proposed trigger event took place so recently, it has been suggested that it may have produced observable effects on Earth. Indeed, David Rubincam (NASA Goddard Spaceflight Center) has recently suggested, in the September 25, 2015, online issue of the journal *Icarus*, that a conspicuous spike in the extraterrestrial ^{3}He abundance recorded in deep ocean sediments at the Eocene-Oligocene transition boundary might potentially be a consequence of the Gould Belt formation event. Likewise, other researchers have suggested that the Gould Belt formation event may have gravitationally perturbed the Oort Cloud, inducing a stream of cometary nuclei to move into the inner Solar System and thereby enhance the terrestrial impact rate.

Indeed, in the time frame of interest the Chesapeake Bay and Popigai impact structures were formed. These particular impact craters are 85 km and 100 km in diameter, respectively, and both are estimated to be some 35 million years old. The reality of any linkages, causes, and effects between these terrestrial phenomena, cometary impacts, and the events associated with the formation of Gould's Belt (given that it is indeed a real and coherent structure) remains open to debate, but it is an active area of research.

The nearest giant molecular cloud at the present epoch, TMC-1, is found in the direction of the constellation of Taurus, situated on the sky about mid-way between El Nath (β Tauri) and the Pleiades and almost directly above Aldebaran (α Tauri). It is not an especially obvious structure to the naked eye, other than invoking the general sense of a less star-studded region of the sky. The dust clouds associated with TMC-1 were first described by British astronomer John Hind in 1852. Indeed, it was at this time he identified the reflection nebula (NGC 1555—often called Hind's Nebula) associated with the star T Tauri, to which we shall return to shortly. Being so close to the Sun, TMC-1 spans about 10 degrees of the sky, and although it is a stellar nursery it does not contain any conspicuous HII regions, such as that found in the Orion Nebula. Figure 4.18 shows a composite view of TMC-1 made at multiple infrared wavelengths. Here the filamentary structure and the dense, star forming cores of the cloud are revealed.

The variable star T Tauri, associated with the nebula NGC 1555 (Fig. 4.19), as first identified by Hinds, is a remarkable object, and following the pioneering studies by Alfred Joy (Mount Wilson Observatory) and Victor Ambartsumian (Burakan Observatory, Armenia) in the 1940s and 1950s it has become the prototypical example of all newly forming low-mass stars. This being said, T Tauri has a mass about twice that of the Sun and is a member of a multiple star system [6].

Located about 360 light years away, T Tauri is seen in the same region of the sky as the Hyades cluster, although it is not a Hyades cluster member. With an estimated age of

Fig. 4.18 The Taurus molecular cloud, TMC-1. Distinctive filamentary structures of gas and dust are visible in this 3-color ($\lambda = 160$, 250 & 500 μm) image based on data collected with the Herschel (infrared) Space Observatory. Image courtesy of ESA/Herschel/PACS

Fig. 4.19 Hind's Nebula (NGC 1555) associated with the T Tauri star system [6]. Image courtesy of 2MASS/NASA

smaller than half-a-million years, T Tauri is in the final stages of becoming a genuine star. It is still moving towards its main sequence state and the onset of hydrogen fusion reactions within its central core (see Chap. 5).

Although it is still accreting material from its natal cloud, via a circumstellar disk, T Tauri is also losing material via a pair of bipolar jets—the jets constrained to flow away from the disk in a direction aligned with the spin axis of the star. The mass outflow of T Tauri amounts to some 10^{-7}–10^{-8} M$_\odot$/year—a mass loss rate that is 5 million times higher than that of the Sun through the solar wind. The variability of T Tauri stars in general is driven by erratic variations in the accretion rate and by the presence of large starspots—the latter being a consequence of the typical rapid rotation (of order two to three times faster than that of the Sun) and the presence of strong surface magnetic fields. The young ages, typically less than a million years, that can be associated with T Tauri stars makes them a good tracer of star formation environments.

Figure 4.20 shows the result of combining a CO emission survey map with the positions of T Tauri stars in the TMC-1 molecular cloud. Not only are the T Tauri stars invariably associated with regions of high gas density, indicative of the idea that stars form in the densest, clumpiest regions of giant molecular clouds, the T Tauri stars must also

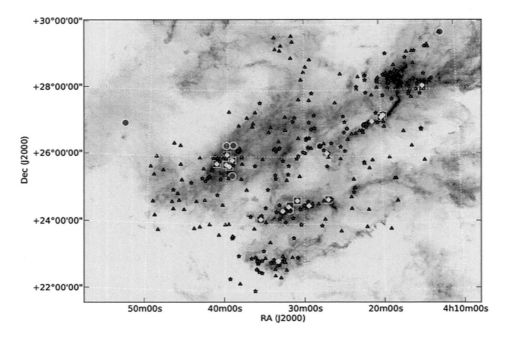

Fig. 4.20 CO emission survey map of TMC-1 with superimposed locations of T Tauri stars (*small dots and triangles*). The position of various T Tauri stars supporting outflows are shown as *yellow diamonds, red triangles, and green dots*. Image courtesy of FCRAO

deposit a considerable amount of energy into the surrounding molecular cloud. In a 2012 study by Gopal Narayanan (University of Massachusetts) and coworkers published in the journal of the Royal Astronomical Society, some 20 T Tauri stars sporting outflows of gas were identified in the 100-square degree survey of TMC-1 by the Five Colleges Radio Astronomy Observatory (FCRAO). The topic of gas motion within star-forming clouds, and the process of star formation itself, will be picked-up again in Chap. 5.

Born of Stardust

All atoms, other than hydrogen and helium, are born in the ultra-high temperature, high density cores of stars and in supernova explosions, but molecules must form in much lower temperature and substantially more quiescent environments. Additionally, once formed, molecules are prone to dissociation through interactions with UV photons, and accordingly if they are to survive for very long they need shielding from high energy radiation. All of these conditions, high gas density, relative quiescence, and UV shielding are achieved within molecular clouds, and the key enabling mechanism for the production of interstellar molecules is interstellar dust.

The basic building blocks of molecules are, of course, atoms, and in terms of cosmic abundance measures, at the present epoch, the six most common atoms are hydrogen, helium, oxygen, carbon, neon, and nitrogen. Helium and neon, both being noble gases, shun all interactions with more plebian atoms and play no role in active chemistry, so in terms of the construction of interstellar molecules the main ingredients (in decreasing abundance) are going to be H, O, C, and N.

Based on pure numbers, hydrogen dominates, and spectroscopic observations reveal a cosmic abundance of elements such that for every 10 million hydrogen atoms there are something like 7000 oxygen atoms, 3000 carbon atoms, and 900 nitrogen atoms. Since H atoms are so abundant, it is expected that hydrogenation reactions with C, N, and O will be common, leading to the production of such molecules as water (H_2O), ammonia (NH_3), and methane (CH_4). And indeed, these molecules are observed within molecular cloud structures.

Carbon monoxide, CO, so important for the mapping of molecular clouds at radio wavelengths, is produced efficiently within the warm gas phase of the interstellar medium via the reaction chain: $O + H_2 \rightarrow OH + H$, followed by $C + OH \rightarrow CO + H$. In cold gas regions CO can be formed via what are called ion-molecule interactions, beginning with $C^+ + OH \rightarrow CO^+ + H$, with the eventual formation of HCO^+, which then combines with an electron to produce $CO + H$. The important mechanism for generating ions, such as the C^+, in the interstellar medium and molecular clouds, is cosmic ray (CR) interactions, where $CR + C \rightarrow C^+ + e^- + CR'$, with CR' being the slightly less energetic cosmic ray after the ionization event.

Of major importance for our specific narrative is the formation of molecular hydrogen, H_2, since it is the dominant interstellar molecule. It turns out that H_2 production can precede in a number of different ways. In the gas phase direct radiative attachment can take place such that $H + H \rightarrow H_2 + \gamma$, where the photon γ carries away the excess 4.5 eV binding energy of the newly formed H_2 molecule. So-called associative detachment can also produce H_2 in the gas phase through the reaction $H + e^- \rightarrow H^- + hf$, followed by $H^- + H \rightarrow H_2 + e^-$. It turns out, however, that the production of H_2 in the gas phase is rarely a very efficient process, and accordingly a third growth mechanism has been invoked, and this appears to be the major channel for the formation of not just H_2 but essentially all complex interstellar molecules. Initially described in detail by (Sir) William McCrea and Derek McNally in 1960, this mechanism works according to a hit-and-stick method, whereby H atoms (and, indeed, any other incident atoms of O, C, and N) collide with and then accumulate on the surface of dust grains. Rather than being static once captured, the H atoms can migrate over the grain surface and on encountering another atom (most likely another H atom) start to synthesize molecules. Accordingly, as time progresses, a patina of molecules is built up on a grain's surface, with the molecules being ejected into space as a result of ongoing collisions with other atoms or the grain entering a higher temperature environment.

Of great importance for the origin (and eventual sustenance) of life (at least on Earth) is the production of water—a molecule composed of two hydrogen atoms and one oxygen atom. A water molecule can grow via the gas phase and the grain-surface accumulation mechanism, where reactions such as $O + H \rightarrow OH$, $OH + H \rightarrow H_2O$, or $OH + H_2 \rightarrow H_2O + H$ can operate.

Although there is still much to be unraveled and understood about interstellar chemistry, it appears that most of the oxygen in the interstellar medium, if not already captured into CO or contained within refractory silicates, will eventually end up as water-ice, and much of this water-ice will be found in the form of icy mantles that envelope dust grains. The mantle ices will undergo additional processing through their exposure to ultraviolet radiation and cosmic ray bombardment.

Laboratory experiments reveal, for example, that UV irradiation can produce a whole range of complex organic molecules, two such examples being the formation of formamide (NH_2CHO) and ethanol (C_2H_5OH). For temperatures between 275 and 483 K formamide is a liquid, and it has been proposed as an alternate solvent to water for supporting life, although this would require a different biochemistry from that found on Earth. Ethanol (more commonly known as alcohol) is also a liquid, and this potent fluid has sustained life, in one form or another, on Earth for many centuries. In regions of intense star formation the grain mantles will become heated and begin to sublimate, leading to an enhanced abundance of gas phase molecules. It is in this rich gas phase environment that larger organic molecules can begin to form.

Once formed and residing in either the gas phase or on the surface of an interstellar grain, a molecule is still prone to disruption. Indeed, while UV radiation and cosmic rays can aid in the formation of molecules, they can also result in their disruption. Photodissociation,[8] via such reactions as $hf + H_2 \rightarrow H + H$, and $hf + OH \rightarrow O + H$, is always going to be important within GMCs and especially so in the outer, less shielded regions. Likewise, dissociative recombination will also always occur, via reactions such

---

[8] Here the photon has energy $E = hf$, where h is Planck's constant and f is the frequency.

as $e^- + OH^+ \rightarrow O + H$, within molecular cloud regions. Ion molecular reactions can also work to break molecules apart, and reactions such as $He^+ + CO \rightarrow C^+ + O + He$ will act to reduce the CO abundance.[9] To build up a molecular cloud, therefore, the rate of producing molecules must exceed that at which they are destroyed, and this is where the shielding effects of interstellar grains, high gas density, and long physical column depths all come into play. The topic of GMC formation and disruption will be discussed in the next chapter.

Although the grain-surface catalyst mechanism is highly important for the growth of interstellar molecules, the question now arises as to where the grains themselves come from. For this we need to look for hotter and higher density regions than those found in the interstellar medium. Indeed, interstellar grains are formed in the outer envelopes of aged stars just prior to and during their terminal planetary nebula or supernova phase. In such high density, relatively warm environments, C and O atoms can directly condense out of the gas phase to make refractory structures—just like soot grains condense out of a candle flame. The grains, therefore, grow by condensation and accretion growth in the cooling outflows from old stars, and they are then launched by radiation pressure into the surrounding interstellar medium.

According to the predominance of O or C in the envelope of the aging star, so silicates (e.g., $SiO_2 =$ silica) or carbon-rich compounds (e.g., amorphous carbon, graphite, diamond, and polycyclic aromatic hydrocarbons = PAHs) will form. Supernovae, in contrast, can build up grains containing iron, sulfur, aluminum, and magnesium, as well as silicates. A careful comparison between the abundance of elements such as Si, Fe, Mg, Al, and Ca in the interstellar medium reveals that they are significantly under abundant with respect to their Solar System values. This observation indicates that the great majority of such elements must be incorporated into grain structures. Indeed, it is estimated that of order 45 % of Si atoms, 75 % of Fe atoms, and 70 % of Mg, Al, Ca, Ti, and Ni atoms are locked up within interstellar grains.

The picture that has emerged with respect to the structure of interstellar dust grains is a complex one (Fig. 4.21). The core of each grain is composed of silicates, iron, and carbon compounds of one form or another. Recalling the discussion relating to the observed polarization of starlight and the mapping out of the galactic magnetic field, it is additionally required that the core, in general, be elongated. It is the preferential magnetic field alignment of spindle-shaped grains, recall, that produces the starlight polarizing condition. Surrounding the core is a mantle of various ices, including H_2O, CO_2, and NH_3 ices. Moving within and upon the surfaces of the grain mantles are numerous atoms and molecules that interact at random to produce still more complex molecules. The molecules so produced are then ejected into space either as a result of the formation of the molecule itself or by the heating of the dust grain along with the sublimation of the surrounding ice mantle.

---

[9] The ionized He^+ atom in this reaction is produced via a cosmic ray interaction with a neutral helium atom: $CR + He \rightarrow He^+ + e^- + CR'$.

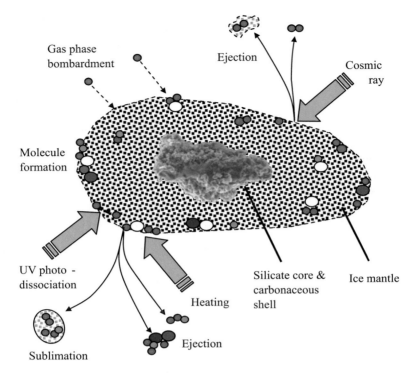

Fig. 4.21 Schematic picture of an interstellar dust grain. The core is composed of silicates, iron, and carbonaceous elements, while the enveloping mantle is composed of ices and embedded atoms and molecules. Gas phase, cosmic ray, and UV photon bombardment as well as heating result in the formation of new molecules on and in the ice mantle as well as the ejection and disruption of molecules

The Pillars of Creation

And on the peak a fiery Eagle:
It had a throat spewing fire
A mouth pouring flame
Feathers as a fire whirling
And as sparks sparkling.

—Elias Lonnrot, *A Perilous Journey – The Kalevala* (1835)

It is said that a picture paints a thousand words, and it could be added that a picture also spawns myriad thoughts and possibilities. Some pictures, however, take us beyond the ability of words to describe, and they pass beyond an ill defined boundary to become iconic and definitive. Such images convey more than textual and visual information. They stimulate the nerves directly, they make the mind and body tingle with excitement, and they fix an epoch.

The history of astronomy is well supplied with iconic images, and such images have, over the centuries, inspired generations of lay readers and practicing astronomers alike. Indeed, iconic images provide humanity with those all-important Hernán Cortés moments, as eulogized by John Keats, standing on that barren peak in Darien seeing the sweeping vista of the Pacific Ocean for the very first time—a view that was entirely new, vast, deep, enthralling and full of possibility. Such images and vistas change the way in which we see the world.

In recent times the Hubble Space Telescope (HST) has supplied an abundance of incredible portraits of the universe, and among the very best of its "snapshots" is the iconic Pillars of Creation (Fig. 4.22), vast columns of interstellar gas and dust stretching through

Fig. 4.22 The Pillars of Creation. This HST image was captured by Jeff Hester and Paul Scowen (Arizona State University) on April 1, 1995. It is a composite of 32 separate exposures made at different wavelengths of light with the Wide Field and the Planetary Camera aboard the HST. The color coding in the final image is such that the light emitted by hydrogen appears *green*, that of ionized sulfur *red*, and that of doubly ionized oxygen as *blue*. Image courtesy of NASA/ESA/HST

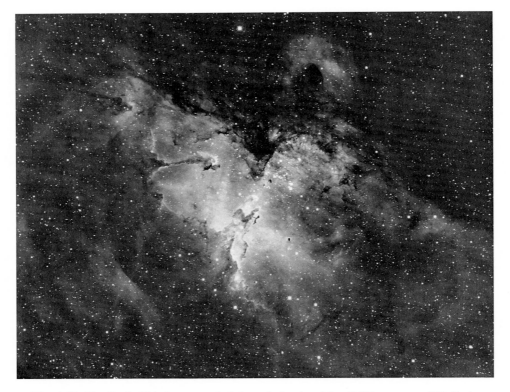

Fig. 4.23 The Eagle Nebula as recoded through the 4-m Mayall Telescope on Kitt Peak. Image courtesy of the National Optical Astronomy Observatory

the cold depths of interstellar space, the same unfathomable abyss that once so terrified Blasé Pascal but now fascinates modern society. The image name is apt, and it carries gravitas. The Pillars are part of the Eagle Nebula, and although they look more like a shadow-boxing caricature of Ishirŏ Honda's Godzilla, they are actually the eagle's talons (Fig. 4.23). The Spire, or less dramatically, column V, provides a second set of grasping talons for the eagle (Fig. 4.24).

The images of the Eagle Nebula grab the eye and hook the brain. There is a rugged and desolate beauty to its composition. It is otherworldly, bedazzling, dark, and in many ways foreboding—the eagle's talons are reaching out to snatch at us. We are its hapless prey. The eagle's head and beak, like a tilting headstone, is arched downward, its wings outspread with pinions curving.

The Eagle Nebula resides in the constellation of Serpens Cauda, "the Serpents body," and the faint glow from of its constituent dusting of stars was first recorded by Swiss astronomer Jean-Philippe de Chéseaux in a catalog of star clusters produced in 1746. Presenting the results of his study to the Académe Française des Sciences, Chéseaux described nebulae number 4 in his list as simply being 'a star cluster,' and to this he added a set of coordinates in the sky.

Fig. 4.24 The Spire, an isolated companion to the Pillars of Creation. Image courtesy of NASA/ESA/HST

Neither the nebulae nor its associated stars are actually visible to the unaided naked eye. The eagle flies beyond our natural ken, and at this early stage in the narrative no common name was applied to 'nebula number 4.' Present-day observers indicate that with a modest, 6-inch telescope, perhaps some 10–20 stars can be seen in M16. With a 10-inch

Fig. 4.25 The open cluster NGC 6611—the powerhouse at the heart of M16. Numerous Bok globules are visible near the bottom of the frame. Image courtesy of NASA/HST

telescope and a dark, moonless sky, however, the faint green-grayish sheen of the nebula can be glimpsed in the eyepiece. Appearing like the faint smudge of a smoke-ring billowing on the heavens, the nebulosity appears against a dark backdrop peppered by numerous closely grouped stars (Fig. 4.25).

Charles Messier, the great comet sleuth of the eighteenth century, rediscovered Chéseaux's nebula in June of 1764 and for the first time specifically mentions that the stars appeared to be "enmeshed in a faint glow." As an object not to be confused with a comet, Messier stored the nebula as number 16 in his now famous catalog. It was a nebulosity to be ignored when sweeping the sky for new cometary interlopers. Messier provided an estimate for the apparent size of M16 in the sky, giving it a diameter of some 8 minutes of arc. Modern tables record the size of M16 as being some three times larger than Messier's original value, and the *Observer's Handbook* of the Royal Astronomical Society of Canada indicates that it covers an area in the sky more like 30 by 20 minutes of arc in extent.

The closest reasonably bright star to M16, as seen in the sky, is ν-Ophiuchi, also known as Sinistra (Latin for the left side), and the nebula sits just a few degrees away from the star gamma-Scuti (Fig. 4.26). Situated, as it is, almost exactly on the galactic plane and located just north of the constellation Sagittarius, M16 is cradled in the arch of the Milky Way, and it is surrounded by a rich field of globular clusters, and this, as we shall see, indicates that our eye, when seeking out M16, is looking towards the center of the Milky Way.

In addition to its more historical designation of M16, the Eagle Nebula is also cataloged as NGC 6611 in John Dreyer's 1888 *New General Catalogue of Nebulae and Clusters of Stars*, where it is described as being a "cluster of at least 100 large [that is, bright] and small stars." In Dreyer's later and revised *Index Catalog of Nebulae and Clusters of Stars* (published in 1898, with further corrections in 1912) the nebula is cataloged as IC 4703 and described as being "bright, extremely large, cluster M16 involved."

Although the name Eagle Nebula is familiar to present-day astronomers, it is an appellation given only in recent times. It is not clear who first coined the evocative name for the nebula, and it may well have circulated in astronomical vernacular long before being used in print. An Ngram[10] review of the 13 million books digitized by Google Books, however, indicates that the name Eagle Nebula first saw common usage beginning in the 1970s, with the word count frequency of 'Eagle Nebula' apparently overtaking that of 'Messier 16' in the 1990s. Linkage of the term Pillars of Creation with the Eagle Nebula additionally began in the mid-1990s, and the latter is seemingly a direct result of the press release associated with the now iconic HST image (Fig. 4.22).

As dramatically revealed in Fig. 4.23, the Eagle nebula is an active star formation region located within a much larger molecular cloud structure. Based on studies of the stars within NGC 6611, the cluster and nebula are situated about 2000 pc away within the Sagittarius-Carina spiral arm. The light we see from the Eagle Nebula today took flight some 6500 years ago, at a time when the ancient Egyptians were first beginning to build their pyramids and the ancient Britons were constructing early versions of Stonehenge. Eagle Nebula starlight has taken all of recorded human history to reach us.

The stars located within NGC 6611 (Fig. 4.25) represent the power house of the Eagle Nebula, and the cluster contains many massive stars, varying between 2 and 85 $M_\odot$. The cluster also supports a large population of newly forming stars, and it is estimated to be just 2–3 million years old [7]. The cluster also contains many thousands of already formed stars and has an estimated total mass of 10,000 solar masses, but it is the most massive stars within the cluster that are responsible for the ionization and animation of the Eagle Nebula, producing a vast HII bubble (the body of the eagle) that is some 200 light years across.

It is also the massive, most luminous stars of NGC 6611 that have been photo-evaporating the parent molecular cloud (the Eagle's nest) and sculpting the Pillars of Creation. The Eagle Nebula contains at least five massive dusty pillars, or, as astronomers

---

[10] See: http://books.google.com/ngrams.

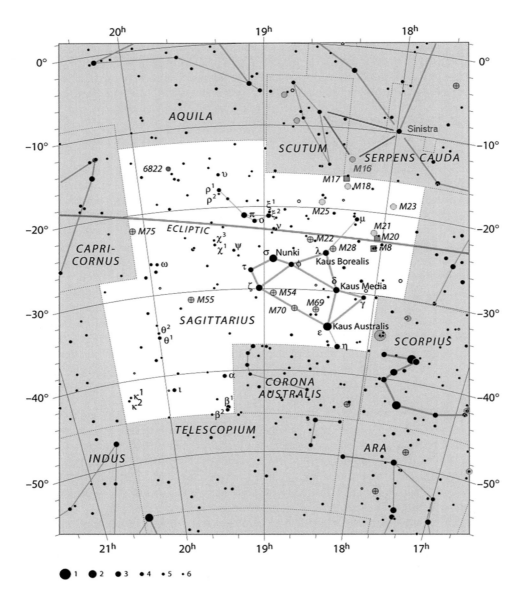

Fig. 4.26 Star map showing the location of M16. Forming an equilateral triangle (*red lines*) with the stars ν-Oph (Sinistra) and α-Scuti, M16 sits at the southernmost vertex. In the diagram the *filled circles* correspond to open clusters (some of which have Messier catalog numbers), while the *filled circles with a cross* represent the positions of globular clusters. The *squares* correspond to various bright nebulae. Image courtesy of IAU/Sky and Telescope

have called them in other star forming nebulae, 'elephant trunk' features. The Pillars of Creation image (Fig. 4.22) reveals Pillars I, II and III, with Pillar IV being located just out of the bottom of the frame; the Spire (Fig. 4.24) is the V column.

The Pillars are extensive structures, being many light years in length, and their shape is determined by a photo-evaporation driven wind that is pushing lower density material away from dense molecular cloud cores. The dense cores at the tip of each pillar essentially act as a shield that forces the ionization flow into a finger-like (elephant trunk-like) protuberance that points away from the ionizing source. In a sense, the pillars are comet-like tails, with the tail pointing away from the luminous stars in NGC 6611. The three-dimensional structure of the pillars is shown in Fig. 4.27. The tip of the left-most pillar in Fig. 4.27 (column I) is behind NGC 6611 and is pointing towards our line of sight. The other three pillars to the right (II, III and IV) have their tips pointing away from us, and they correspondingly appear less bright.

The ionization fronts that are sculpting and shaping the pillars have also resulted in the formation of numerous evaporating gaseous globules (EGGs). These dense knots of gas and dust are typically a few hundred astronomical units in size, and it is thought that they indicate the locations of active star formation. Indeed, infrared imaging of the Pillars (Fig. 4.28) reveals that most of the molecular material is located at the tips of the pillars and within the EGGs, and appropriately, within the EGGs are found the clear signatures of young stellar objects (to be discussed in Chap. 5), which are literally stars in the making.

Fig. 4.27 The three-dimensional structure of the Pillars of Creation. The location of NGC 6611 is towards the *upper right* here, and the pillars trail away from the most luminous stars in the cluster, like cosmic versions of outdoor windsocks. Image courtesy of ESO/M. Kommesser

Fig. 4.28 Near infrared image of the Pillars of Creation. At the longer wavelengths of infra-red light the dust obscuration that hampers optical observations is less severe, and the interior structure of the pillars, or more precisely the lack of structure, is revealed. Image courtesy of ESO/M. McCaughrean & M. Andersen

Additional infrared imaging of the Eagle Nebula, made with the Spitzer space telescope, reveals that the cavity of the HII region, making up the body of the Eagle, is full of warm dust (Fig. 4.29), and this opens up the strong possibility of past supernova activity within NGC 6611.

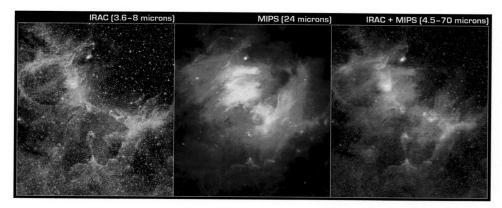

Fig. 4.29 Spitzer spacecraft infrared images of the Eagle Nebula. The *central* image reveals the warm dust filling the HII cavity of the nebula. The *right-most* image reveals the contrast between the hot (supernova heated) dust, shown in *green*, and the cooler surrounding dust of the star-forming pillars. The Pillars of Creation are located to the *middle right* in each image, while the Spire is located at *bottom* in the center. Image courtesy NASA/JPL-Caltech

The Eagle Nebula, along with the Omega Nebula (M17—see Fig. 4.26), are both extensive star formation regions located within the Sagittarius-Carina spiral arm, and CO emission maps reveal that there is a tenuous chain of molecular gas running between the two nebulae. Indeed, Yoshiaki Sofue (University of Tokyo) and coworkers, in a 1986 study of the region, described the bridging CO feature as forming a "long string of beads." Sofue et al. also argued that the 'string of beads' represents a sequential chain of star formation in which the stellar winds and supernovae-blown bubbles, first produced in the cluster NGC 6604 (age 4 million years), have expanded outward to trigger star formation in M16 (age 2 million years) and then M17 (age 1 million years).

The Cat's Paw Nebula (NGC 6334), located about 1.7 kpc from the Sun, affords us another example of a long, drawn-out, pillar-like star formation complex (Fig. 4.30). Indeed, a recent (2016) study of NGC 6334 by Philippe Andre (Université Paris Diderot) and coworkers indicates that it is some 10 pc long and includes at least two massive star forming sub-regions. The authors also determine from their infrared survey data that the nebula has a line mass that varies between 500 and 2000 $M_\odot$/pc, with an associated molecular hydrogen column density of order 10^{28} m^{-2}. Unlike the columns in the Eagle Nebula, which are being sculpted by photo-ionization (recall Fig. 4.27), the extended, strand-like structure of the Cat's Paw Nebula is thought to be constrained and sculpted by the compression of multiple expanding HII bubbles.

The most luminous star formation complex so far studied by astronomers, simply known as W51, also shows a long, strand-like morphology (Fig. 4.31). Located in the Sagittarius spiral arm W51 is about 5 kpc from the Sun, and it is actively forming massive stars and contains in excess of 700 young stellar objects in the mass range between 5 and 20 $M_\odot$. The visual extinction in the direction of W51 is extremely high and estimated to be in excess of $A_V \sim 25$, and accordingly it can only be deep-probed at infrared and radio wavelengths. In terms of contributed energy output, however, it has been estimated by James Urquhart (Max Planck Institut für Rasdioastronomie) and coworkers, in a 2014

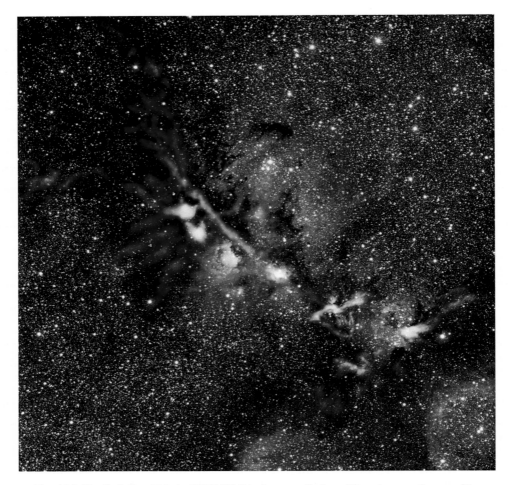

Fig. 4.30 The Cat's Paw Nebula (NGC 6334) in the constellation of Scorpius was discovered by John Herschel in 1837 and is one of the closest star formation regions to the Sun. This image reveals the dust emission at a wavelength of 0.35 mm. Image courtesy of the ArTeMiS team/ESO

paper published in the *Monthly Notices of the Royal Astronomical Society*, that W51 contributes nearly 7 % of the total energy output of all of the embedded massive stars in all of the star formation regions in the entire galaxy. This one star formation complex, accordingly, generates an energy output equivalent to about 5.0×10^6 $L_{\odot}$.

The 2014 study by Urquhart et al. was specifically concerned with mapping out the galactic distribution of massive star formation regions, and, to this end, the authors conducted a massive survey effort, combining radio telescope observations of molecular clouds (CO, CS and NH_3 emission line surveys), infrared survey data (such as the 2MASS survey), as well as visual observations, and data gathered with the Midcourse Space Experiment (MSX) satellite. Their quest was to quantify the position, distance, and properties of as many massive young stellar objects (MYSOs) and ultra-compact (UC) HII regions.

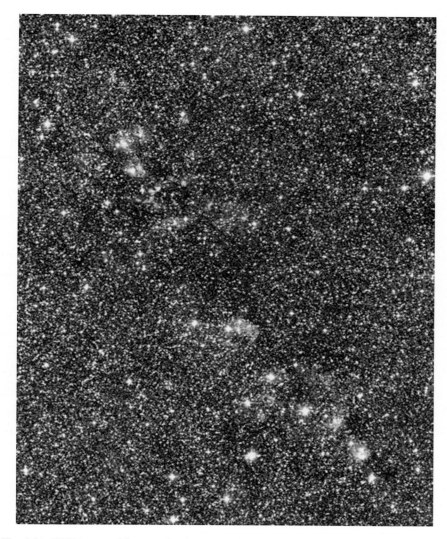

Fig. 4.31 W51 is one of the most luminous star-forming complexes known. Located 5.4 kpc away the region is actively forming massive stars at the present epoch. The linear, strand-like distribution of ultra-compact HII regions is clearly evident in the image, running from the *upper-left to lower right*. Image courtesy of the 2MASS survey, University of Massachusetts

Published as part of the Red MSX (RMS) survey, hosted at the University of Leeds in England, Urquhart et al. were able to identify nearly 1750 MYSOs and UC HII regions, out to distances in excess of 15 kpc from the Sun and across the disk of the galaxy (Fig. 4.32). Since the survey is restricted to the identification of those regions forming massive stars, their distribution will accordingly map out the spiral arm structure of the galaxy—that is, the youngest star formation regions at the present epoch. Figure 4.32

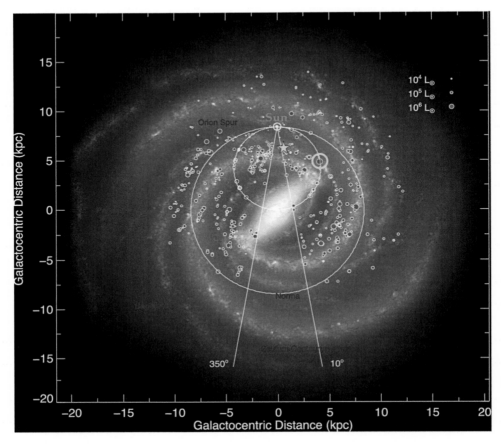

Fig. 4.32 Spiral arm structure of the Milky Way Galaxy as traced out by the distribution of newly forming massive stars and ultra-compact HII regions. The location of the Eagle Nebula is shown by a *yellow star symbol*, and the most luminous known star-forming region W51 is *circled*. The *large white circle* indicates the path of the Sun around the galactic center. The central galactic region between the longitude lines labeled 10° and 350° was not included in the RMS survey. Image courtesy of J. S. Urquhart, Max Planck Institut für Rasdioastronomie, the RMS survey team and the Spitzer-Science Center

shows the results from the RMS study, and it is seen that the newly forming massive stars, with ages less than a few millions of years, trace out a pattern consistent with a four-armed spiral pattern (compare with Fig. 2.24). Also shown in Fig. 4.32 is a sketch of the galactic core region, indicating that it is an elongated, bar-like structure some 10 kpc in length (Indeed, the bar runs along a diagonal of the 5 kpc molecular ring).

The spiral arm features do not run to the very center of the galaxy; rather they are attached to the ends of the central bar. This is in contrast to the Whirlpool and Andromeda galaxies (Figs. 3.1 and 3.2), where the spiral arms merge into a spherical central bulge [8]. That our galaxy has a central, elongated bar-like distribution of stars, angled at about 45°

to our line of sight to the galactic center (Sagittarius A*) was long-suspected, but essentially confirmed through survey data collected with the Spitzer (infrared) Space Telescope in the first decade of this century.

The Spitzer Space Telescope data is more sensitive to the detection of cooler, lower mass stars, and in addition to picking out the galactic bar structure, the survey data revealed the existence of just two spiral arm features. This seemingly contradictory result is, in fact, simply related to a difference in the ages of the stellar populations being studied. The RMS survey is concerned with massive stars, and these objects, because of their relatively short lifetimes, will be seen where they formed. The Spitzer data, however, is preferentially looking at cooler, longer-lived, low-mass stars, and accordingly these objects have had a chance to spread out within the galactic disk, piling-up, in effect, into two distinctive arm-like concentrations bounded between the arms by the more massive stars.

Are GMCs Dangerous?

In the heavens, as on Earth, we live in dangerous times, and we know it. Not just the every-day dangers of simply crossing the road, but the existential dangers of a comet or asteroid impact [9]. Thankfully for us, existential dangers are very rare, and the odds of such events occurring in any given lifetime are astronomically small. The Solar System, however, is 4.5 billion years old, and when it comes to very rare events, time is the great paymaster. How safe, therefore, is the Solar System against complete disruption through the close encounter of a giant molecular cloud? There are multiple aspects to this problem, but we will first consider a stability condition, and then look at the emersion problem.

To estimate how often an encounter between our Solar System and another star might take place we need to set a closest encounter distance d, estimate the number of stars that there are per unit volume of space $n*$, and determine the typical stellar encounter speed $V*$ [10]. Setting the distance estimate $d=1$ pc, the approximate distance to Proxima Centauri at the present epoch, and taking a typical stellar encounter velocity of 25 km/s, the typical encounter time will be of order 160,000 years.

Encounters with stars at 1 pc are fairly common, therefore, and the Solar System will have undergone nearly 30,000 such encounters since it formed 4.56 billion years ago. Setting $d=0.25$ pc, which corresponds to a star skimming the outer boundary of the Oort cometary cloud, the encounter time increases to some 6.4 million years, and perhaps 700 such encounters have taken place over the age of the Solar System. For a stellar encounter as close as, say, 500 AU to the Sun, the encounter time is about 27 billion years, or 6 times longer than the present age of the Solar System. Although close stellar encounters, at the distance of a few pc, are going to be relatively common, a very close encounter that might potentially perturb the orbits of the outermost planets is not likely to occur during the entire 10 billion year main sequence lifetime of the Sun.

Given that very close stellar encounters with the Solar System are extremely rare, what about encounters with molecular clouds? For the molecular clouds we need to modify the encounter time formula [10] for the characteristic number density of molecular clouds. Here we use Fig. 4.17 as our guide; this figure indicates that there are 21 molecular clouds within the cylinder of radius 1 kpc and height 100 pc centered on the Sun.

Accordingly, $n_{mc} \approx 7 \times 10^{-8}$ pc^{-3}, which is about 1.3 million times smaller than the number density derived for the stars in the solar neighborhood. Substituting for n_{mc} and taking a characteristic encounter velocity of 15 km/s, at a close encounter distance of 10 pc, we find an encounter time of 3.4 billion years. For a close encounter distance of 20 pc, the encounter time is reduced to about 860 million years.

From the close encounter times just derived we can infer that the Solar System has undergone perhaps a handful of relatively close encounters with massive molecular clouds. In terms of planetary stability, clearly none of these past encounters has been cataclysmic, since the Solar System is still here. For all their mass, therefore, molecular clouds do not have a great deal of gravitational influence for planets tightly bound to their parent star.

In terms of Solar System stability, at issue during an encounter with a molecular cloud is the difference between the gravitational attraction of the cloud across the diameter of the Solar System, and the gravitational attraction of the Sun. Specifically it is the tidal effect that needs to be determined [11], and even for the most massive of giant molecular clouds, it turns out, this effect is incredibly small. We appear to be safe, at least from this existential threat. The tidal interaction, however, while comforting on the issue of planetary stability, reveals a mystery. The mystery relates to the present-day existence of long-period comets. These comets are traced back to an origin within the Oort Cloud that surrounds the Solar System, stretching out to distances of order 100,000 AU.

Since the Oort Cloud is a highly dynamic structure, with its outer boundary continually adjusting according to the distribution of nearby stars, galactic tides, and the passage of molecular clouds, it supplies a steady flux of comets into the inner Solar System. Indeed, there is the gentle and steady rain of long-period comets that round the Sun, with about 600 such objects being detected per century. But here's the problem. The tidal effects on the Oort Cloud caused by a passing molecular cloud is significant. On this basis the Oort Cloud should have been decimated long, long ago. And that means that it would be a rather exceptional turn of events that long-period comets should still be observable at the present epoch.

Astronomers are not agreed on how to solve this issue. One suggestion, largely developed by Victor Clube and William Napier in the 1980s, suggests that while the Oort Cloud is destroyed during a close molecular cloud encounter, it is also reformed at the same time by a capture process. The argument in the latter case is built on the idea that cometary nuclei can actually form in the cores of molecular clouds. Under this scenario the Oort Cloud may have been destroyed and reformed multiple times since the Solar System formed. Not all astronomers agree, however, with this scenario, and one of the particularly problematic issues with the destroy-and-capture scenario is that no comet with a clear interstellar origin has ever been observed. Indeed, such a comet would be instantly recognizable by its highly hyperbolic orbit. No such interstellar comet has ever been recorded, although, according to how one adopts representative numbers, such a comet should be observed once every 50–100 years. There is absolutely no reason to believe that interstellar comets don't exist, but we are still in a holding-pattern with respect to the first detection.

Given that the Solar System must have encountered molecular clouds in the past it is natural to ask if Earth might show any evidence of these events. The idea that Earth's climate might be modulated through external forcing by astronomical phenomena has a long history, but in terms of recent climate change it is generally accepted that the various ice age epochs during at least the past several millions of years are due to periodic variations in Earth's orbital eccentricity and semi-major axis, combined with periodic variations in the angle of the obliquity of the ecliptic.

These external forcing factors operate on a timescale of order hundreds of thousands of years and were first identified by James Croll in the mid-1860s and the theory further developed by Milutin Melankovic in the 1920s [12]. Other, longer timescales are, however, possible, and additional astronomical forcing factors have now been identified. Harlow Shapley (Mount Wilson Observatory) appears to have been one of the first researchers to suggest that terrestrial climate change might be modulated by passage through interstellar clouds. Writing in *The Journal of Geology* for September 1921, Shapley suggests that interstellar cloud passage must have occurred in the past, and that such encounters could result in long-term changes in the Sun's luminosity, and this would then have a direct affect on Earth's climate.

Unfortunately, Shapley's argument relied on the now rejected idea that galactic novae were produced when a star passed through a dense interstellar nebula—a process driven by either "collision or friction." Accordingly Shapley suggested that if the Sun chanced to pass through a more diffuse interstellar cloud then its luminosity would increase for perhaps a few millions of years, and this "would sufficiently alter terrestrial temperature to bring on or remove an ice sheet." This same basic idea was developed by Fred Hoyle and Raymond Lyttleton in 1939. Writing in the *Proceedings of the Cambridge Philosophical Society*, Hoyle and Lyttleton addressed the accretion issue of interstellar gas directly, finding that the amount of mass gained by the Sun depended directly on the density of the interstellar cloud and the inverse cube of the velocity with which the Sun moved through the cloud. Again, as Shapley had before them, Hoyle and Lyttleton argued for a prolonged increase in the Sun's luminosity due to the accretion of nebula gas and further suggested that this could produce the conditions necessary for the eventual onset of an ice age on Earth. They also suggested that in the more extreme cases of accretion of interstellar gas by the Sun Earth's climate might be driven towards a particularly hot state—such as that exhibited during the Carboniferous epoch between 300 and 350 million years ago.

Here the situation lay for some 30 plus years, only to be picked up again by Sir William McCrea (University of Sussex) in 1975. McCrea started where Hoyle and Lyttleton left off, arguing, just as they had, that during its passage through an interstellar cloud the Sun's luminosity would increase, and this would push, according to some unspecified mechanism, Earth's climate towards an ice age state. McCrea's interesting addition to the argument, however, was that rather than being a totally random process, driven by the Solar System's chance encounter with an interstellar cloud, ice ages on Earth would be inevitable and quasi-periodic. The modulating mechanism invoked by McCrea was that of spiral arm crossings.

The point, McCrea noted, was that molecular clouds and regions of dense interstellar clouds are preferentially located near to and within spiral arms, and that the Solar System, moving at a relative speed of order 10 km/s to the galactic spiral density wave, must pass through a spiral arm once every 100 million years or so. Additionally, the time for the Solar System to traverse a given spiral arm feature will be of order 10 million years. Although McCrea realized the importance of spiral arm crossings with respect to the enhanced likelihood of the Sun and Solar System encountering an interstellar cloud, he pushed the argument too far for most researchers when he wrote, "when the Solar System enters a compression lane [the inner edge of a spiral arm] – there must occur on Earth an ice epoch." It was McCrea's use of the word "must" that made other researchers uncomfortable. Indeed, McCrea was essentially arguing that the ice age onset mechanism was completely removed from Earth (that is, from its active geology, meteorology, and orbit) and entirely due to a spiral-arm modulation mechanism working upon the Sun's luminosity.

Neither Shapley, Hoyle, and Lyttelton nor McCrea were able to identify any specific mechanism by which Earth's climate might be pushed into an ice age by the increased luminosity of the Sun as it accreted material while advancing through an interstellar cloud. This changed, however, in 1976 when in response to McCrea's paper Mitchell Begelman and Martin Rees (both of them at the Institute of Astronomy at Cambridge University) published an article in the journal *Nature* in which they focused more closely on the solar wind rather than material accretion onto the Sun. Begelman and Rees essentially dismissed the idea that the Sun could effectively accrete enough material to appreciably change its luminosity during an interstellar cloud crossing event (and this remains the present day consensus). What they did note, however, is that at those times when the Solar System is pushing through an interstellar cloud, the termination location where the solar wind pushes into the surrounding cloud could be pushed much closer to the Sun than Earth's orbit. At normal times the boundary of the heliosphere is located at about 100 AU from the Sun. With only a modest gas density being required of the intercepting interstellar cloud, Begelman and Rees were able to show that Earth, with its orbital radius of 1 AU, could effectively be excluded from the Sun's otherwise protective heliosphere.

This exposure of Earth to the surrounding interstellar cloud would then open up the atmosphere to an enhanced cosmic ray flux, and it could additionally result in the direct accumulation of interstellar dust in Earth's atmosphere. With these two processes at play the potential for directed climate change was, in principal, very real. Cosmic ray exposure, for example, has been identified as a possible mechanism for cloud seeding, and accordingly, the greater the cosmic ray flux, the greater the global cloud cover, the greater the reflection of sunlight back into space, and the cooler Earth's climate. This mechanism and sequence, however, remains controversial, and not all researchers are convinced of its efficacy. Likewise, the more dust there is in Earth's atmosphere, the greater the amount of sunlight that is reflected back into space, and once again this tends to drive the overall temperature downwards.

The latter mechanism is just the interstellar dust version of the cooling effects due to massive volcanic eruptions. The eruption of Mount Pinatubo in 1991, for example, resulted in a measurable global cooling of about 0.5 degrees in the time interval 1991–1993, and in the same time frame the depletion rate of the ozone layer increased substantially. The latter effect opened up the associated biological hazard of an enhanced UV flux at Earth's surface.

In addition to the high possibility of encountering an interstellar cloud during a spiral arm crossing event David Clark, along with McCrea and Francis Stephenson, noted in a 1977 research paper published in the journal *Nature* that the Solar System would also experience a greater chance of encountering a nearby supernova. The argument here is that since the spiral arms are delineated by regions of star formation, and stars more massive than about eight times that of the Sun will undergo supernova disruption in less than 10 million years, this is the same amount of time that it will typically take the Solar System to traverse a spiral arm.

The possibility of Earth's atmosphere being directly affected by a supernova blast wave, in addition to its brief outburst flash of gamma rays, is enhanced. The gamma rays emitted from a nearby supernova, it is argued, will result in the enhanced, if not total, destruction of the ozone layer, and this will lay the biosphere open to an enhanced UV radiation flux and an enhanced cosmic ray flux.

A nearby supernova explosion would likely have both climactic and biological effects that should be traceable in the geological and fossil records, but in spite of much research effort, no clearly identifiable link to either dramatic climate change and/or extinction events has been established [13]. In recent times it has been questioned whether there is any correlation between dramatic variations in Earth's climate record, biotic extinction events, and the times of spiral arm crossing. In 2009, for example, Andrew Overholt (University of Kansas), Adrian Melott, and Martin Pohl conducted a detailed study of crossing times using an up-to-date model of galactic spiral arm structure. They were specifically interested in the apparent 140-million-year modulation in Earth's climate based upon ^{18}O abundance measurements in fossils. Here the argument being developed is that the measured ^{18}O abundance is a reflection of the ^{18}O to ^{16}O isotope ratio in ocean water. In other words, ^{18}O will be enhanced over the normal ^{16}O abundance during cold epochs, with the ^{16}O being preferentially evaporated from the oceans and then stored in ice sheets. There is no known terrestrial mechanism that can modulate the ^{18}O abundance on the requisite timescale, but it is to order of magnitude close to the expected spiral arm crossing timescale.

The model calculations by Overholt et al., however, found no correlation between the times of ice age epochs and spiral arm crossing events over the past billion years. Additionally, Overholt and coworkers note that the spiral arm pattern within the Milky Way Galaxy is not perfectly symmetric (recall Fig. 2.24), and accordingly spiral arm passage events cannot produce a distinctly periodic climate forcing effect on Earth. Figure 4.33 shows the path of the Solar System around the galactic center and the spiral arm pattern of the Milky Way as deduced from the high-density CO molecular cloud survey data. The last four spiral arm passage events took place about 30, 250, 330, and 400 million years ago, when the Solar System crossed the Sagittarius-Carina, Norma (twice) and the Perseus spiral arms, respectively. Although these new results do not rule out a connection between climate change forcing events and spiral arm crossing times, they do rule out the notion that all ice age epochs are the result of an external astronomical forcing factor.

The mass extinction event at the end of the cretaceous epoch, where the dinosaurs on land and the last ammonite species in the sea became extinct is generally attributed to a large asteroid impact in what is now the Yucatan peninsula in Mexico. The evidence for this impact is the massive Chicxulub impact crater, which has been assigned an age of 66 million years. Evidence for the importance of this impact is not only found in the crater itself but in the presence of an iridium (Ir) abundance spike in the global end-Cretaceous boundary clay layer. This spike was attributed, by Luis and Walter Alvarez and coworkers in 1980, to the fallout of Ir-enriched material contained within the asteroid being blasted into Earth's atmosphere after the impact.

During the past decade a growing body of evidence indicates that the mass extinction at the end-Cretaceous was not solely driven by the Chicxulub impact. Indeed, there appears to have been multiple large asteroid impacts around 65 million years ago, along with massive geological upheavals associated with the Deccan Traps eruptions in India. Likewise, the climate at this time was undergoing a distinct cooling trend. This cooling trend may

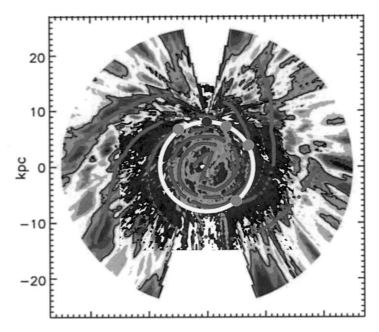

Fig. 4.33 Path of the Sun (*white circle*) about the galactic center with respect to the spiral arm structure of the Milky Way disk. The *blue circle* indicates the present location of the Sun, while the *green circles* indicate the positions of spiral arm passages. The background image shows a combination of the 21-cm neutral hydrogen distribution and the CO distribution (*thick red lines*) in the galactic plain. Image adapted from Englemaier et al.: arxiv:0812.3491

well have been a response to the input of volcanic gases over an extended time interval of some 30,000 years, but it may have additionally been triggered by the passage of the Solar System through a dense ($n \sim 10^{-6}$ H atoms per m^3) molecular cloud core. This latter model was proposed and developed by Tokuhiro Nimura (Japan Spaceguard Association) and coworkers in a paper submitted to the journal *Gondwana Research* in 2016.

The cooling mechanism invoked by Nimura et al. is that of dust accretion by Earth's atmosphere as the Solar System passes through a ~100 pc-sized molecular cloud over a timespan of some 10 million years. While the suggestion and cooling mechanism outlined by Nimura and coworkers is not new, what is particularly novel and interesting is that they do have some backup evidence for the argument. Specifically, they have looked at the iridium levels associated with several deep-sea core samples from the Pacific Ocean floor. These cores date from the time interval 78 to 10 million years ago and show a broad iridium anomaly extending from 72 to 64 million years ago. The distinct Ir spike, due to the Chicxulub impact, 65 million years ago, is also seen in the core data.

What Nimura et al. focus on, however, is the broad 8 million-year-long iridium anomaly, which cannot possibly be due to a single asteroid impact. This is where the slow and steady accretion of dust from a molecular cloud passage event comes into play. The cooling associated with cosmic dust accretion is not so much invoked as the main driver of the end-Cretaceous mass extinction, but it would certainly be another important stress factor on the biosphere. Additionally, Nimura and coworkers note that at the time of close encounter with a molecular cloud there is likely to be strong perturbation effect on bodies in the outer Solar System, the Oort Cloud specifically, resulting in a high influx of comets to the inner Solar System and a concomitant increase in the terrestrial impact hazard. The arguments, egos, and theories accounting for the end-Cretaceous extinction will no doubt continue to run for many decades yet, but it is now abundantly clear that it was not just one event that resulted in the sad loss of the dinosaurs[11] but probably a whole combination of terrestrial and astronomical factors.

Although there is no clear evidence in the geological, fossil, or climate record to indicate that Earth has ever suffered any truly catastrophic effects from a nearby supernova explosion, this is not to say that traces of past nearby supernova activity cannot be found. Again, to find evidence for recent events one needs to go to sea and extract deep sediment samples. Most recently this has been studied by two research groups, one led by Dieter Breitschwerdt (Berlin Institute of Technology) and the other by Anton Wallner (Australian national University). Both articles appeared in the April 7, 2016, issue of the journal *Nature*.

Although Wallner and coworkers studied deep-sea sediment data, Breitschwerdt's group looked at the transport properties of supernova blast waves. The key supernova marker that Wallner et al. looked for in their sediment cores was the radionuclide ^{60}Fe. This radionuclide has a half-life of about 100 million years, and it is only known to form during supernovae explosion events. Accordingly, if any sediment sample contains active ^{60}Fe, it must have been deposited within the ocean during the last several hundred million years. The specific deposition time can be gauged by measuring the abundance of ^{60}Fe along with that of its decay product ^{60}Ni.

By finding the age of the sample sediment (via ^{26}Al and ^{10}Be radionuclide analysis) and measuring the ^{60}Fe abundance Wallner and colleagues found evidence for Earth having swept up supernova ejecta material at two past epochs—one between 1.5 and 3.2 million years ago and the other between 6.5 and 8.7 million years ago. No extensive mass extinctions have been noted at these past geological times, but both epochs are associated with times during which the global temperature decreased. Wallner et al. concluded that the ^{60}Fe data indicates that Earth has encountered debris and ejecta from a supernova explosion, at a stand-off distance of 100 pc or smaller, once every 2–4 million years.

The study by Breitschwerdt's group complements that by Wallner et al. and looked to model the historic supernova record of the local interstellar medium. Specifically, Breitschwerdt and co-workers argue that the Local Bubble was likely carved out by some 14–20 supernova explosions within the past 13 million years by a moving group of massive stars, the surviving lower mass members of which are now in the Scorpius-Centaurus stellar association (recall Fig. 4.17 showing the nearby molecular cloud distribution). They also conclude that one or two supernovae will have occurred within 60–130 pc of the Sun from 1.5 to 2.6 million years ago.

---

[11] Of course, the extinction of the dinosaurs was good for mammals, of which humanity is a member. It is still rather sad, however, that such wonderful beasts as the *Stegosaurus* and *Diplodocus* are no longer with us on Earth.

The galaxy is a dynamic, explosive, and potentially dangerous place, but our present existence and the analysis presented here indicates that in terms of existential disasters we need not overly fear close stellar encounters, nearby supernova explosions, or encounters with giant molecular clouds. Nonetheless, as we shall see in the following chapter, we are largely made of atoms that were forged in the central cauldrons of stars and supernovae. We need not fear what goes on in the galaxy, but we do owe our very existence to the processes that operate within the interstellar medium.

Life in a Molecular Cloud

For many observers the stars in the night sky are a visual mystery, their arrangement being a pointillist composition of random lights with names and groupings abstracted from ancient Arabic phrases and mythical Hellenistic beasts. But all observers are at the least aware of the stars. This latter understanding seems to go almost without saying, but it need not always have been the case.

Great science fiction writer Isaac Asimov speculated on this very topic in one of his classic short stories, written in 1941, called "Nightfall." This particular story is set on an imagined world situated within a sextuplet star system. The remarkable orbital ballet performed by the system's host stars results in the habitable world, named Lagash (a name invoking a dry and heat-blasted domain), being continually bathed in sunlight—except, that is, for one brief nighttime interval that marches around every 2500 years. For the unfortunate generation that suffers the time of darkness it is a moment of impending fear and uncertainty; to them the circumstances of the last nightfall are a matter of mythology and ancient history, and no one alive knows, for sure, what will happen. Worst of all, however, there is a rumor, a ridiculous rumor, that the darkened sky becomes full of stars— myriad sparkling lights, just like the familiar six, but cold, distant, and remote—uncountable stars embedded within the vastness of infinity.

Asimov's story begins with a company of astronomers (a body of scientists restricted to a universe of just six stars—imagine that!) attempting to make a prediction for the exact onset time of nightfall. The story ends with the howls and unbelieving screams of people driven mad by the unfolding view, a view that is entirely unremarkable and obvious to us, of the great vault of the heavens. For us the idea of infinity is exemplified by the stars, but if the Solar System, by pure chance or bad luck, had happened to stray into a dense molecular cloud region, say 10,000 years ago, then we, too, could be as hapless of the stars, beyond our singular Sun, as the imagined inhabitants of Lagash.

What would be required of an enveloping interstellar dust cloud to make the Sun's present-day stellar neighbors invisible to the human eye? For the night sky to be rendered a stellar void, sporting us a view of just the Sun, the Moon, the planets, and the occasional wayward comet, the necessary extinction would amount to that which would render the star Sirius, presently the brightest star in the sky, as faint as, say, visual magnitude +7. With Sirius thus obscured, all other stars, even the six closer ones to the Sun, would be invisible to the naked eye.

For Sirius the important details are its distance, found by parallax observations to be $d=2.65$ pc, and its absolute magnitude (see the Appendix in this book), which is deduced to be $M=+1.47$. With these quantities so determined the apparent magnitude comes out as $m=-1.43$. In order to push the apparent magnitude of Sirius above $m=+6.5$, the threshold for human vision, the total extinction required in our calculation is $A=7.93$. The characteristics of the Solar System enshrouding cloud are now made clear [14], and the total number of particles within our hypothetical veil amounts to a staggering 6.4×10^{48} dust grains. Indeed, if these grains were all brought together and packed into a massive cube, then the cube would have sides some 3 million km in length.

Our imagined dust cube has sides that are 300 times larger than Earth, but only 3 times larger than the Sun (see Fig. 4.34). While our compacted cube of hypothetical dust seems large by human standards, the actual dust cloud, when the grains are evenly spread out within a sphere having a radius equivalent to that of the distance to Sirius, is actually quite unremarkable. The density of dust grains within a typical dark nebula can vary from 10^{10} to 10^{15} grains per cubic meter, and accordingly our hypothetical dust cloud is a mere wisp of a cloud compared to what nature can offer. Indeed, the number density of grains required to completely blot out all the stars from the sky is only 1000 times greater than that deduced for the Local Bubble (recall Fig. 2.11). As human beings we are indeed decidedly fortunate that the stars can be seen at all.

Apart from rendering the celestial vault starless, what else might be changed if the Solar System was enshrouded by a dense dust cloud? The answer to this is pretty much all of human history. With no celestial sphere to act as a reference system, perhaps only the lunar calendar would have been opened up to development, and the seasons, although just as regular as they are now, would be much more difficult to gauge since there would be no helical rising or setting markers to indicate their progress.

The year could still be determined, of course, by keeping track of the Sun's rise and set locations with respect to local horizon markers. The planets would still be visible in a dust-enshrouded Solar System, but the development of an understanding of planetary orbits

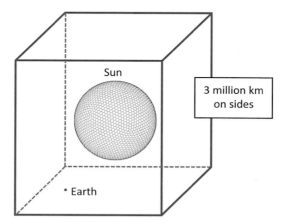

Fig. 4.34 A 3-dimensional scale diagram comparing the size of Earth, the Sun, and the imagined compacted dust cloud that would render the sky completely starless from Earth, if uniformly dispersed within a sphere having a radius equal to the distance of Sirius

would have been much more difficult to establish—with no celestial reference frame only relative planet on planet positions could be measured, and such details as retrograde motion would have been much more difficult to unravel. The fundamental observations and achievements of Brahe, Copernicus, and Kepler would likely have been much delayed beyond their actual times. Indeed, without Kepler's laws and a celestial reference frame for measurements Newton's gravitational theory may have taken much longer to develop. And, while comets would still be observable in a dust-enshrouded Solar System, without stellar reference points it would be next to impossible to determine their orbits, and accordingly Edmund Halley's great prediction, relating to their occasional periodic return, would never have occurred when it did.

Although it would have taken longer to understand the orbits of comets in a starless night sky, it is highly probable that the planet Uranus would have been known to the ancients, and so, too, would the asteroids Vesta and Juno, these latter three objects hovering on the threshold of naked-eye visibility. There is no specific reason to change the time in human history at which the telescope, that quintessential tool of the modern-day astronomer, was invented, and it may well be that the pace of discovery of Solar System objects would have been no different to that which we know. The Jovian moons and the phases of Venus and the mountains on the Moon would all have been within the reach of the telescopes used by Galileo and his contemporaries. Galactic astronomy would presumably have been somewhat slower to develop in a dust-enshrouded Solar System, since the isotropic dust distribution would obscure the faint glow of the Milky Way from naked-eye and low power telescopic view.

Likewise, the idea of their being separate and independent galaxies would probably take much longer be to develop, such studies requiring the development of large and expensive telescopes, and with no obvious objects for them to look at, the funding for such speculative projects would presumably be hard to win—indeed, even harder than it is in the modern era. Einstein's revolutionary development of general relativity may well have been developed as our history timeline indicates, but its experimental proof would not have been possible. Without a celestial reference frame to make observations against, the anomalously high advancement in the perihelion of Mercury may not have been known by 1915, and certainly the famous 1919 eclipse demonstration of the gravitational bending of starlight would not have been possible.

If the stars were not accessible at optical wavelengths to test general relativity, then the first experimental verification of Einstein's theory would have been delayed by some 40 years, awaiting the results of the 1959 Pound-Rebka gravitational redshift experiment carried out at the Jefferson Laboratory at Harvard University. This laboratory, rather than an astronomy-based experiment, set out to measure the time-dilation effect predicted by Einstein's theory to apply across regions of varying gravitational field strength.

One of the many famous sayings of the great (and sadly fictional) detective Sherlock Holmes runs along the lines that any true philosopher should be able to infer the existence of Niagara Falls from the observation of a single drop of water, and accordingly, a daring theoretician could in principle infer the existence of other stars even if they were not visible. To do this, however, the theoretician would need all of the analytic power of modern-day electron microscopes and spectrum analysis. The existence of other stars would accordingly be buried within the micro-structure of meteorites and/or cometary dust grains. It is within these miniscule time capsules that the pre-history of the Solar System is betrayed, and a careful analysis of oxygen isotope ratios (and a detailed knowledge of stellar—that is solar—evolution) will reveal that some of the grains and micro-diamonds buried deep within meteorite fragments must have formed around other stars (recall Fig. 3.15). Likewise, of course, the existence of stars and galaxies and the interstellar medium would have been revealed to human consciousness once radio, infrared, and microwave wavelength telescopes and detectors were produced, but this would bring us right in to the 1970s timeframe of our specific historic timeline (recall Fig. 2.16).

Not only would the history of astronomy (and the sciences in general) timeline of discovery be totally different if we chanced to live in a dust-enshrouded Solar System, so, too, would that of human exploration and navigation. With no North Star it would be next to impossible for navigators, away from local landmarks, to judge their latitude, and without a celestial framework the determination of longitude, by the lunar method, for example, would not be possible. The moons of Jupiter, chronometer, and solar noon time methods of position determination would still be open to navigators, but no specific and always accessible nighttime methods would be available to the hapless mariner. How such limitations on navigation would have changed world exploration, conquest, and trade development is hard to imagine, but one suspects that events would have played out at a much slower rate than our actual history portrays.

Not only would human history have been much different if the Solar System had chanced to reside within a long-lived dust region, so, too, might that of evolution. Many creatures use the Moon phase, Earth's magnetic field, and seasonal temperature variations to regulate their life cycles and behaviors, but others use stellar observations. Many migrating birds, for example, use stellar reference markers at nighttime during their annual migration. And, remarkably, the humble dung beetle (*Carabaeus Lamarck*) is known to use the faint glow of the Milky Way as a reference line from which to navigate by.

As alternative histories go, the one in which humanity is denied the sight of stars in the night sky is perhaps the most alien to us. It is difficult, even unsettling, to imagine a night sky devoid of stars, although many modern-day city dwellers have almost achieved this unholy situation. The stars of the night sky have inspired great works of literature and poetry, for indeed, where would humanity be without the star-crossed lovers so beautifully invoked by Shakespeare, or the stars to cast our wishes upon, and even the wandering stars to hitch our proverbial wagons to. And again, for all of its unmitigated nonsense, without the stars there would be no astrology to guide our days or foretell the future outcomes of our present-day actions. Without the stars, and the vastness of the celestial vault, how would we imagine infinity, and how would we understand the remarkable circumstances of our very existence? Indeed, a history without stars is hardly worth considering ☺.[12]

---

[12] Attempts a literary humor should never be too obscure, but to explain the happy face emotive, the point here is that the word *consider* is derived from the Latin phrase to *consider the stars*. I didn't say it was great literary humor.

REFERENCES

1. In the radio, long-wavelength range the so-called Rayleigh-Jeans law offers a very good approximation to Planck's equation for a blackbody radiator. In this manner the energy flux density varies according to the inverse wavelength squared: $I \sim$ constant/λ^2. Accordingly by moving to a wavelength of 9-cm, Reber expected an increase in the signal intensity, with all else being the same, of $(14.6/0.09)^2 = 26{,}316$.

2. Baade and Minkowski had a famous bet concerning the energy source relating to Cygnus A. Baade originally challenged Minkowski to a prize of one-thousand dollars, but later reduced it to a bottle of whisky, that Cygnus A was the end product of the collision between two galaxies. Baade claimed the prize when Minkowski was able to detect emission lines in the optical spectrum of Cygnus A with the 200-inch Mt. Palomar telescope, and Minkowski conceded. Baade complained that his prize from Minkowski was only a hip flask of whisky, rather than a full bottle. It also transpires that during a visit to Baade's office a few days after the bet was settled, it was Minkowski who emptied the hip flask. This, in fact, is as it should be since it is now known that Baade was wrong and that the colliding galaxies model is incorrect. The idea of an accretion scenario took many years to develop, but it essentially grew out of a research paper published in 1963 by Fred Hoyle and William Fowler. Rather than invoke a black hole, however, Hoyle and Fowler suggested that, "a large stellar-type object of large mass" resided at the galactic center. The stars considered were truly massive, having masses in the range 10^5 to 10^8 $M_\odot$, which radiated at the maximum possible value – the so-called Eddington limit. The role of the central superstar was to generate both the relativistic electrons, via interactions with a surrounding disk of interstellar medium, and then trap the electrons within the host stars magnetic field, and thereby produce the observed synchrotron radiation. The present day model for the generation of jets and radio emission simply swaps out the superstar for an accreting supermassive black hole.

3. Not only can a galactic civilization intentionally broadcast its existence to the galaxy, it may also broadcast its presence by radio communication unintentionally. Radio broadcast signals have been leaking from the Earth for well over a century and this means that we are surrounded by a radio bubble having a radius of some 100 light years (946 trillion meters). From the number density of stars (derived in chapter 1) this volume of space will contain more than 10,000 stars. Humanity has also broadcast its existence directly, the first detailed broadcast being sent from the Arecibo radio telescope in 1974. The total broadcast time for the message was just 3 minutes, and the wavelength adopted was $\lambda = 12.6$-cm.

4. The term water hole was introduced by Barney Oliver in 1971 who additionally suggested that the search for deliberately beamed signals could proceed between wavelengths of 18 and 21-cm. The 18-cm wavelength corresponds to a strong emission line associated with the OH hydroxyl molecule. Oliver's idea was that as water (H_2O) appears to be a fundamental requirement for the existence of life (at least as we currently know it), so the combination of HO and H gives H_2O, and this sets the region of the water hole at radio wavelengths.

5. A Maser (microwave amplification by stimulated emission of radiation) works through the stimulated, rather than spontaneous emission of light. To produce a maser a gas must contain molecules that have been placed into a population inversion state. That is, the number of electrons in high energy states is greater than would be expected under the conditions of thermal equilibrium.

6. The optical primary of T Tauri, designated T Tau N, is a spectral type K0 star, while the secondary, T Tau S, is a close binary composed of an M spectral type secondary (T Tau Sb) and an as yet unclassified primary companion (T Tau Sa) – remarkably, it appears that T Tauri Sa has a mass of 2.1 $M_\odot$, which is comparable to that of T Tau N, but most of its atmospheric emission is blocked in our observational line of sight by a surrounding edge-on disk of gas and dust. In contrast, it is though that T Tau N is being viewed pole on. While T Tauri is the accepted protype for a protostellar system it is really a close dynamical group of three stars undergoing accretion, and there are still many uncertainties about its past (and even present) activity. Indeed, the great Otto Struve identified a new nebulous cloud (designated NGC 1554 = Struve's Lost Nebula) in association with T Tauri in 1868, but within 10 years of its first detection the cloud had disappeared. A more recent 2016 study of the T Tauri system by Markus Kasper (ESO) and co-workers (arXiv:1606.00024v1) has revealed new features in the star's circumstellar-gas environment. In addition to several new reflection nebulae features, Kasper et al. interestingly find a new and, "previously unknown coiling structure …. in reflected light". The origins behind this feature are presently unclear, but it is possibly related to a precessing jet of out-flowing material associated with T Tau N

7. An extremely detailed review of the Eagle Nebula, along with its remarkable pillars and the galactic cluster NGC 6611, is provided by Joana Oliveira (Keele University, England) in *The Handbook of Star Forming Regions Vol II*. Bo Reipurth (Ed.), Astronomical Society of the Pacific (2008).

8. The percentage of barred spiral (SB) galaxies to ordinary spiral (S) galaxies has apparently varied over the age of the universe. Recent studies indicate that in the early universe only about 20% of spiral galaxies exhibit a central bar. Closer to the present epoch, however, the percentage of barred spiral galaxies is of order 70%. This increase in the relative number of barred spirals is thought to reflect the growth time of the bar-generating instability – the more massive a disk, so the more rapidly a bar-like organization of stellar orbits can come about. Indeed, the observations suggest that barred spiral galaxies did not appear as a morphologically distinct type until the universe was at least 6 to 7 billion years old. At earlier times, it is suggested, tidal interactions between galaxies were much more common, and galaxies were still accumulating mass; such conditions acting against the initiation of the bar-forming instability. Material is actually flowing along a bar towards the galactic center, rather than away from it and into the spiral arms, and this material fuels on-going star formation in the galactic bulge. There is evidence to suggest that galactic bar structures need not be permanent, and that spiral galaxies can change between barred and non-barred configurations over time – the transformation time from one state to another being of order several billion years.

9. The author has discussed such existential dangers and their associated timescales in the book *Rejuvenating the Sun and Avoiding Other Global Catastrophes* (Springer, New York, 2008).

10. The typical time interval between star encounters is given by the expression $T_{encouter} = 1/(\pi d^2 n* V*)$, where d is the closest approach distance, $n*$ is the number of stars per unit volume of space (recall from chapter 1 that $n* = 0.09$ stars/pc^3 in the solar neighborhood), and $V*$ is the encounter velocity. Casting the distance in units of parsecs and the velocity in units of 10 km/s, so the encounter time becomes $T_{encouter} = 4\times10^5/(d^2 V_{10}*)$ years.

11. If we take the diameter of the solar system to be d_{ss}, then the gravitational attraction at the outer perimeter due to the Sun will be $g_{sun} = G M_{\odot} /(d_{ss}/2)^2$, where G is the universal gravitational constant and $M_{\odot}$ is the mass of the Sun. If the molecular cloud has a mass M_{mc}, and the solar system passes as close as a distance d_{mc} from the clouds center (here we assume the cloud is spherical), then the difference in the gravitational field due the molecular cloud across the diameter of the solar system will be $\Delta g_{mc} = G M_{mc}/d_{mc}^2 - G M_{mc}/(d_{mc} + d_{ss})^2 \approx 2 G M_{mc} d_{ss}/d_{mc}^3$. With these two expressions we can calculate the ration $\Delta g_{mc}/g_{sun} \approx \frac{1}{2} (M_{mc}/M_{\odot})(d_{ss}/d_{mc})^3$. While $M_{mc} >> M_{\odot}$ it is the second term, $(d_{ss}/d_{mc})^3$ that dictates a small tidal interaction from even the most massive molecular clouds. Taking $d_{ss} = 100$ AU, for example, and $d_{mc} = 10$pc $= 2.1\times10^6$ AU, so $(d_{ss}/d_{mc})^3 \approx 10^{-13}$, and this factor far outweighs the mass ratio term which might rise as high as 10^6 or 10^7, but no more.

12. The author has discussed the details of the Croll-Melankovich theory in the book *Terraforming: the Creating of Habitable Worlds* (Springer, New York, 2009).

13. For more technical details, see the authors review paper on the topic, The past, present and future supernova threat to Earth's Biosphere, in *Astrophysics and Space Science*, **336**, 287–302 (2011).

14. The required density of dust grains ρ_g must be such $A = 2.5 \log10(e) \tau$, where $\tau = \pi a^2 Ng$, and where a is the dust grain radius (we assume they are all the same size), Ng is the number density of grains along the line of sight, with $Ng = \rho_g d$, and for us d is the distance to Sirius – what happens beyond the distance to Sirius is not our direct concern. Indeed, even if, miraculously, the dust density dropped to zero beyond the distance to Sirius, the damage would have already been done, and from the Earth no stars will be seen by the unaided human eye. Substituting in the values for A and d, and taking the dust grains to have a characteristic radius of $a = 10^{-7}$ m, we find that $\rho_g \sim 3 \times10^{-3}$ grains per cubic meter. With this density we can calculate the total mass of grains within our assumed spherical dust cloud stretching out to Sirius – the mass of grains turns out to be some 5.4×10^{31} kg, which is about 1/27th of the mass of the Sun, or equivalent to some 28,265 Jupiter mass planets.

5

In the Grip of Gravity

The stars, from whence? – Ask Chaos – he can tell.
These bright temptations to idolatry,
From darkness, and confusion, took their birth.

—Edward Young, the poem "Night Thoughts," written between 1742 and 1745

Gravity is the weakest of all the fundamental forces, but for all of its want of strength its reach is very long, and it dictates the large-scale structure of the galaxy. Indeed, it is the action of gravity that controls the fate of matter contained within molecular clouds, and it is the long reach of gravity that makes star formation possible. Before looking at the details of star formation, however, let us first briefly summarize what has been deduced about the interstellar medium in earlier chapters, and to again contradict Walt Whitman let us do it in numerical and tabular form. Table 5.1 presents the characteristic numbers and identifiers of the various phases of the interstellar medium.

Bringing the characteristic numbers together reveals a steady increase in number density, a decrease in temperature, and a decrease in size when moving from the high volume filling coronal gas to the very low volume filling but mass dominating molecular clouds. Additionally, as we saw in Chap. 2, H II regions pinpoint star-forming nurseries, and the observations further indicate that all star-forming nurseries are contained with molecular cloud clumps and cores.

Indeed, molecular clouds are the pillars of stellar creation, and remarkably, in order to make stars, nature has first to create regions that have temperatures nearly as low as they can possibly be.[1] It is within the frigid zones of molecular cloud cores that stars are conceived, since it is there that gravity can sink its proverbial teeth into the slow-moving molecules and begin to produce collapsing structures that develop larger and larger central densities. In contrast, a hot gas cloud of fast-moving atoms and molecules will tend to

---

[1] The coldest known natural object is that of the Boomerang Nebula (NGC 40). Observations with the Atacama Large Millimeter/submillimeter Array (ALMA) telescope in 2013 reveal that this protoplanetary nebula has regions where the gas temperature is ~1 K (about 1° above absolute zero). These regions are even colder than the 3 K cosmic microwave background (CMB) radiation.

© Springer International Publishing AG 2017
M. Beech, *The Pillars of Creation*, Springer Praxis Books, DOI 10.1007/978-3-319-48775-5_5

Table 5.1 The various phases of the interstellar medium in descending order of associated temperature

Phase	Temperature (K)	Number density (m^{-3})	Characteristic size (pc)	Observations
Coronal gas	10^6	10^4–10^2	galactic (50%)	UV & X-ray
Warm neutral	10^4	5×10^5	galactic (15%)	21-cm radio
Warm ionized	10^4	5×10^5	galactic (25%)	21-cm radio
H II	10^4	10^8–10^{10}	1–10 (<1%)	Optical
Cold neutral	100	5×10^7	galactic (5%)	21-cm radio
GMC	15	10^8	45 (<1%)	IR & radio
MC	10	5×10^6	10	IR & radio
MC clump	10	10^9	5	IR & radio
MC core	10	10^{11}	0.25	IR & radio

In Column 4 the term "galactic" implies that these phases are widespread across the galactic disk, and the bracketed terms indicate a characteristic volume-filling factor

GMC giant molecular cloud, *MC* molecular cloud

dissipate before gravity can corral the components into clumped and collapsing higher density structures. The march to stardom, therefore, is all about steadily increasing the density of material in a given small volume of cold interstellar space and then letting gravity work its magic.

It is estimated that there are something like 4000 GMCs within the disk of our galaxy, and that the typical separation distance between such clouds is of order 500 pc. Figure 5.1 shows what the GMC distribution within the disk of our galaxy would look like to an observer located in a distant galaxy. This figure reveals the spiral arm structure of the Milky Way, and it is based on the catalog of 1064 molecular clouds observed within the galactic disk and published in the May 1, 2016, issue of *The Astrophysical Journal* by Thomas Rice (University of Michigan) and coworkers. These clouds and their denser sub-structures range in size from a just a fraction of a parsec to of order 40–50 pc across, with each cloud containing from a few tens to perhaps a few million solar masses of molecular gas and dust.

The mass spectrum of molecular clouds indicates that the cumulative number of clouds $N(M)$ having masses greater than M behaves as a power law expansion with $N(M) \sim M^{-\gamma}$, where γ varies between 1.5 and 2.5, and this indicates that most of the molecular mass resides in a relatively few large clouds rather than in numerous smaller ones. Additionally, the recent study of molecular clouds by Rice and coworkers indicates that the power index γ and the upper mass limit vary according to galactic location. In the inner galaxy the power law index is found to be $\gamma = 1.6$, with the upper mass limit being of order 10^7 M$_\odot$; in the outer galaxy $\gamma = 2.2$ and the upper mass limit is 10^6 M$_\odot$. These observations indicate that the inner galaxy is seemingly more efficient at producing larger molecular clouds, although the exact reasons for this are still unclear.

We saw in Chap. 2 that the various low-density phases of the interstellar medium, from the coronal gas to the cold neutral clouds, are in approximate pressure equilibrium (that is, their characteristic temperature multiplied by their characteristic number density is approximately the same). However, the giant molecular clouds along with their substructures are sufficiently massive that they are self-gravitating. Indeed, as will be seen below one of the present mysteries concerning giant molecular clouds is why there are, in fact, any at all, and why it is that their self-gravity hasn't long ago transformed them into stars.

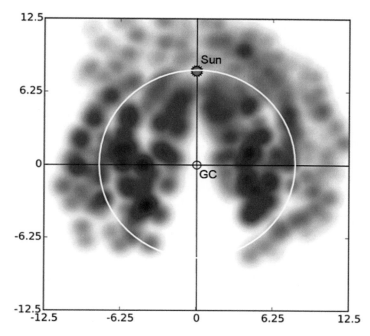

Fig. 5.1 A simulation of what a CO radio telescope survey of the Milky Way Galaxy would look like to an observer situated 70 Mpc away. The square grid has sides corresponding to a scale of 25 kpc, and the Sun is located 8 kpc away from the galactic center (GC). Background image from Rice et al.: arXiv:1602.0279

The Life and Times of a Molecular Cloud

Given a molecular cloud of mass M and initial radius R, what is its characteristic time t_{coll} for collapse under gravity given no opposing forces? This is the so-called free fall, or dynamic timescale, and to order of magnitude it can be determined according to the inverse square root of the density: t_{coll} (years) $\sim 2 \times 10^{-3}/[\rho \ (\text{kg/m}^3)]^{1/2}$, where ρ is the average density in kg per cubic m [1]. For giant molecular cloud structures the gas density is of order 10^{-20}–10^{-19} kg/m³, and this indicates collapse times of order 10^7 years. This is a remarkably short timescale given a Milky Way Galaxy that is at least 12 billion years old. Clearly, something must be stopping molecular clouds from collapsing most of the time. Otherwise, the entire reservoir of interstellar matter would have long ago been converted into stars. This result immediately informs us that star formation must be a very inefficient process.

Indeed, the star formation efficiency ε_{sfr} can be evaluated as: $\varepsilon_{\text{sfr}} = t_{\text{coll}} \ (M_{\text{sfr}}/M_{\text{GMC}})$, where M_{sfr} is the observed star formation rate and M_{GMC} is the total amount of gas contained within molecular clouds. With characteristic numbers of $M_{\text{sfr}} \approx 1.5 \ M_{\odot}$/year (in our galaxy at the present epoch) and $M_{\text{GMC}} = 10^9 \ M_{\odot}$, we have about a 1.5 % star formation efficiency. The deduced efficiency further indicates that something like 100 $M_{\odot}$ of molecular gas needs to be generated within the galaxy per year. The problem to understand with respect to star formation, therefore, is not so much how stars are made (although that is difficult enough), but in finding mechanisms by which the process of star formation within molecular clouds is held at bay.

The answer to this slowing down problem is complex, but it is partly related to the way in which molecular clouds form, partly related to the internal dynamics of the molecular gas, partly related to environment and age, and partly related to the large-scale galactic magnetic field. Let us look at the molecular cloud formation mechanisms first.

Observations of the Milky Way reveal that dense strands of interstellar dust and giant molecular clouds are preferentially located along the leading edges of spiral arm features. Indeed, it is now reasonably clear that the spiral arms of our galaxy are delineated by long filamentary strands of molecular gas. A recent radio telescope study by Catherine Zucker (University of Virginia) and coworkers published in the December 3, 2015, issue of the *Astrophysical Journal* has revealed a skeletal-like structure of long, high-contrast filaments, which Zucker et al. describe as "bones," in the direction of previously recognized spiral arm features. The point, however, is that the continued detection of these bones might ultimately reveal new spiral arm features.

In contrast to the molecular cloud distribution, the brightest, most recently formed stars, along with young stellar clusters, tend to be located towards the trailing edges of spiral arm features (recall Fig. 4.32). These observations are suggestive of the idea that at least some molecular clouds are built up as a result of gas compression driven by a spiral density wave that rotates throughout the galactic disk. The spiral arm crossing time is of order 100 million years, and this further constrains the age of molecular clouds to be 10^8 years $> t_{life} > t_{coll}$. Other mechanisms for the production of molecular clouds invoke either stellar feedback processes or large scale instabilities. In other words, it appears that molecular clouds can be built from the bottom up as well as from the top down. In this manner smaller clouds can coalesce or become swept up to produce bigger clouds, and/or instabilities in the greater galactic disk are enabled to grow and produce massive clouds that then can either remain as they are or fragment into strings of smaller units.

The dominant process in the growth of small molecular clouds is thought to be that of converging gas flows. In this situation stellar feedback compresses the interstellar medium, and this allows for the formation of UV radiation shielded regions where molecules can grow. The stellar feedback is primarily the result of proto-stellar outflows from T Tauri stars, the expansion of H II regions, the strong winds from newly formed massive (O and B spectral type) stars, and supernovae explosions. Additionally, it is the very same feedback mechanism that disrupts the natal molecular cloud out of which the young stars initially formed. In this manner molecular clouds are really very ephemeral structures, sowing the seeds for their disruption at the very same time that they are actually forming stars. It is by forming stars in their interiors that molecular clouds initiate their own destruction. In short, and somewhat ironically, the star formation efficiency of molecular clouds is very low primarily because some stars actually form.

The physical processes responsible for the formation of stars are controlled by the internal characteristics of the molecular clouds, and in essence it all boils down to a competition between gravity, which is always trying to bring material together, and those forces that work against collapse, the latter including such phenomena as rotation, turbulence, and magnetic fields. The interplay of these competing forces will be discussed later.

Before looking at the processes involved in star formation, however, it will be useful, at this time, to take a brief look at what we've learned about stars, and specifically their structure and properties. This knowledge has been painstakingly gleaned over many centuries by asking such straightforward questions as what they are, how they are powered, whether they are all like the Sun, whether the Sun is a star, whether they are all of the same age, and how long do they last. These questions, and many more, have driven the long history of astronomy, but they have mostly been answered in the relatively short, but brilliant, history of astrophysics beginning in the late nineteenth century.

Getting to Know the Stars

There is an old trick-question that asks, "What is the closest star to Earth?" The person to which the question has been posed usually twists their lower jaw and closes one eye, all to the ultimate effect of blurting out something like, "Alpha … No… Proxima Centauri." To which, of course, we all smile and say, "No. The nearest star to Earth is the Sun."

Indeed, the Sun is so familiar to us that we hardly think of it as a star, the latter being lights that are faint, distant, small, and only visible at nighttime. But the Sun is a star, a rather rare star it turns out, but it is a star nonetheless. When, however, did the Sun become a star, and the stars suns? Here is a question that actually has a long history but no clear answer.

Nearly 2500 years ago, the Greek Philosopher Anaxagoras (circa 450 b. c.) suggested that the Sun was a blazing, red-hot rock, larger than the Peloponnesian peninsula, and that the stars were the same, just much further away. Anaxagoras had no specific justification for his statement, but the principal idea, that the Sun and the stars are one and the same kind of object, is at least present. Over time, other philosophers have expressed the exact opposite idea, positing that the Sun is indeed special and very different from the stars. The Sun, after all, is big and hot, the stars are small and provide no obvious heat, and the Sun moves around the sky at a different rate to that of the constellations. What was lacking historically, of course, was data to work with. The unaided eye alone cannot provide the critical information needed to resolve the question.

Tycho Brahe, the last of the great naked-eye observers, posited that the stars are arranged in a shell just beyond the orbit of Saturn, the thought of vast and empty tracts of space existing between the last planet and the stars being taken as an anathema. If so located then the stars must surely be different from the Sun; they must be small, and less bright parochial lights. For all this, however, inspired by the "New Star" of 1572, Brahe commented, "to great applause" according to Kepler in his 1610 *Conversation with the Sidereal Messenger*, that there was ample material within the Milky Way out of which to create new stars.

Galileo, with his newly wrought telescope, writes in his *Siderius Nuncius*, however, that stars cannot be resolved into disks, and hence, if they are like the Sun, then they must be at vast distances from us. Galileo noted that the Milky Way could be resolved into stars, though, and this prompted Kepler to negate his earlier praise of Brahe's notion that new stars might be born within the depths of space. Kepler additionally argued that the Sun must also be special and different from the stars—this latter contention being based mostly on humanity's special place with respect to creation: "In the center of the world is the Sun, heart of the Universe, fountain of light, source of heat, origin of life and cosmic motion." Isaac Newton appeared to favor similarity between the Sun and stars, and he derived a distance estimate to the bright star Sirius that was 615,500 times larger than Earth's orbit about the Sun. Although this number is on the large size (Sirius is actually about 544,500 AU away) it correctly places the stars at a very great distance.

Newton did not speculate upon the possibility of new stars being born in space, but he did allow for stars to be renewed in spirit through divinely supervised cometary impacts. Not until the first reliable stellar parallax measurements presented by Bessel, Henderson, and Struve in the late 1830s did the real distance to the stars become apparent, and indeed, this was at a time when it had already been established, through the spectroscopic work by Fraunhofer (see Chap. 2) in 1817, that while different in their exact appearance the spectra of the Sun and Sirius were at least similar.

The identification that the Sun must be a star and that the stars must be suns was put beyond all reasonable doubt in the 1860s through the spectroscopic work of Huggins and Miller in England, Lewis Rutherfurd in America, and Angelo Seechi in Italy. But even then, it was also apparent that not all stars were exactly sun-like; a whole range of absorption line characteristics were readily observed. Additionally, it was still an open question at the end of the nineteenth century as to exactly where and how the various kinds of nebulae fitted into the grand cosmological scheme of things. Were they independent entities, or did they have some role to play in the formation of new stars?

What was needed was a synthesis, a bringing together of concepts and observations, and the first grand synthesis model was that proposed by Joseph Lockyer in 1890 (see below). Importantly, Lockyer (in part) set about addressing the question which asks whether stars are eternal or ephemeral objects. Indeed, the question arises as to whether all of the stars in the universe are of the same age, having formed in one Genesis-like moment, or did the stars form at different times, in an ongoing generative process. Furthermore, having formed, the next obvious question to ask was whether the physical mechanisms responsible for powering the stars falter in time will lead to their eventual demise.

In order to apply a comprehensive theory to the structure of stars, more than just spectral characteristics are needed. Indeed, to know what is going on inside of a star the physicist needs to know such physical variables as mass, luminosity, radius, surface temperature, and composition. With such numbers in place the astrophysicist can set about looking for patterns and relationships between quantities. The first important step in this direction was taken by Ejnar Hertzsprung in 1911. Combining parallax distance estimates with energy flux measurements one can, and Hertzsprung did, construct a diagram showing the variations in luminosity, which measures the total energy output of a star, with respect to spectral type which, recall, is a proxy for surface temperature (see the Appendix in this book).

In general such a diagram might be expected to reveal a scatter plot, the data points revealing no specific relationship between the two quantities. This, however, is not the case. Stars, it transpires, show a very special set of relationships. American astronomer Henry Russell independently constructed the same diagram as Hertzsprung in 1914, and such so-called Hertzsprung-Russell diagrams (Fig. 5.2) have become the *de facto* testing ground of theory against observation.

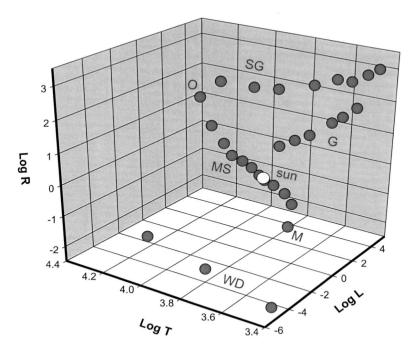

Fig. 5.2 A 3-dimensional version of the HR diagram showing the relationship between luminosity, temperature, and size for stars on the main sequence (MS), stretching from the small, low luminosity, low temperature spectral-type M stars to the relatively large, hot and highly luminous spectral type O stars; the giants (G); the supergiants (SG); and the white dwarf (WD) region. The Sun, a main sequence star, is shown as a *white dot*

The HR diagram is effectively a plot of stellar luminosity versus temperature, and blackbody radiation theory further reveals a link with physical size through the Stefan-Boltzmann law. For historical reasons (see below) the HR diagram plots the temperature as a decreasing sequence to the right along the x-axis (the reverse of normal practice, which would have temperature increasing to the right.) This means that stars having a large radius will plot in the upper right-hand corner of the diagram, and stars with small radii will plot in the lower left-hand corner. In this manner it was deduced that some stars are very luminous, very large, and cool (these are the so-called red giants), while other stars have very low luminosity, very small radii, but high temperatures (these are the white dwarfs) [2].

The vast majority of stars, however, place along a diagonal in the HR diagram that runs from the upper left (high luminosity, high temperature stars) to the lower right (low luminosity, low temperature stars). This prominent diagonal is called the main sequence, and modern-day statistics indicate that 92 % of all stars plot in this region. About 1 % of stars are red giants located above the main sequence, and 7 % are white dwarfs located below the main sequence.

The realization that emerges from the HR diagram is that stars obey specific sets of relationships. The luminosity of a star during its main sequence stage, for example, is not an arbitrary quantity but must correlate with its mass, radius, and temperature. This result was further hammered home by additional studies of binary star systems by Hertzsprung, who was able to show in 1919 that the luminosity of main sequence stars are tightly correlated with their mass. The more massive a star, the more luminous it is. Indeed, a power law relationship between mass M and luminosity L exists for all main sequence stars, with $L = $ constant M^{α}, where $1 < \alpha < 5$.

In contrast to main sequence stars, neither the red giants nor the white dwarfs show any close correlation between mass and luminosity, and of course, the question becomes why. Why are there apparent laws and relationships between M, L, R, and T for main sequence stars (Fig. 5.3) that don't apply to giants and dwarfs, and why indeed, are most stars on the main sequence? Such questions, the direct result of observational inference and measurement, are exactly the sorts of questions that require a physical answer—that is, it falls upon the theorist to explain the observed results according to the established laws of physics.

The rise of astrophysics as a *bona fide* field of study has been nothing short of miraculous and meteoric. It is a field that grew out of developments in late nineteenth century physics, and it specifically grew out of the development of theories relating to thermodynamics and atomic structure. It is the theories relating to heat flow and the properties of hot gases that provide an understanding of the internal structure of stars, and it is the theories of atomic physics that describe the state of matter within a star, and these same theories inform us about the means by which stars can generate internal energy.

As with the majority of new scientific fields of study, there is no one founding figure that can be identified with the rise of astrophysics—there are, however, many distinguished personalities. Such luminaries as (Sir) Arthur Eddington, (Sir) James Jeans, Subrahmanyan Chandrasekhar, Henry Russell, Martin Schwarzschild, Jonathan Lane, and Robert Emden

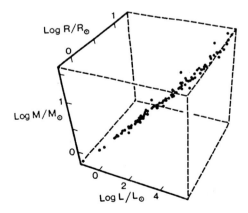

Fig. 5.3 The observed mass-luminosity-radius relationship for main sequence stars. The fact that the stars plot along a diagonal indicates a close relationship between stellar mass, luminosity, and radius. It is the variation in initial mass, set in turn by the star formation process, that determines the run of the luminosity and radius of main sequence stars

were prominent (among many others) in the early development of the theory concerning stellar structure. To make sense of stars, however, at least three branches of study needed to be brought together: the data from observational astronomers, the equations of theoretical physics, and the number-crunching ability of mathematical machinery.

However, these three fields have not always co-existed in harmony. Saliently, pioneering quantum physicist Louis de Broglie was once noted as saying, "Left to itself, theoretical science always tends to rest on its laurels." Theory needs observational constraint, observations need theoretical understanding, and both require the help of mathematical analysis and computation. There is an essential tension that exists between these three fields; the mathematician looks for elegant exactitude, the theorist looks for physical understanding (quite often described by inelegant sets of equations, polynomials, renormalization constants, and various Stradivarius factors[2]), and observational astronomers insist that their results must trump all.

Always one for a good turn of phrase, Arthur Eddington was to write in his 1928 book, *The Nature of the Physical World*, that, "Proof is an ideal before whom the pure mathematician tortures himself. In physics we are generally content to sacrifice before the lesser shrine of plausibility." To this judgment, however, he later added the important point that, "We are all alike stumblingly pursuing an idea beyond our reach." The stumbling pursuits of the physicists eventually, by the mid-1920s, had narrowed down the equations of stellar structure—there are four of them—and these equations describe the run of physical quantities of matter from the center of a star to its surface.

This all sounds rather simple, but there are many associated problems. Knowing the equations to solve is one thing, being able to solve them is quite another. The mathematicians realized, for example, that the equations of stellar structure can only be solved for numerically, and that there is no guarantee that any apparent solution is unique [3]. The physicists knew that they didn't fully understand the state of matter at the high temperatures prevalent in stars, nor did they understand how, or indeed in which part of star, its energy was generated. In spite of such limitations, however, Eddington was able to explain why a mass-luminosity relationship should exist for main sequence stars, and Chandrasekhar was able to use the new quantum physics to explain the properties of white dwarfs.

The understanding of why most stars must reside on the main sequence grew out of the realization that stars are mostly made of hydrogen (70 % by mass), a result first indicated by Cecilia Payne (later Payne-Gaposchkin) in her famed 1925 doctoral thesis,[3] and that it was through the fusion of hydrogen into helium that enabled stars to stay luminous—a process first described in detail by Carl von Weizsäcker and, independently, by Hans Bethe in 1938. The problem as to why there are such objects as red giants in the HR diagram was the last theoretical issue to fall, and the answer lay in the realization that during its main sequence phase a star must develop a chemical discontinuity at its core-envelope boundary.

The first red giant models to be developed along this theme were presented by Estonian astronomer Ernst Öpik in 1939, and soon thereafter more detailed models were developed by Fred Hoyle and Raymond Lyttleton through the 1940s, as well as by Martin

---

[2] These are the various "fiddle" factors that are typically required to make a theory actually *work*.
[3] Payne's thesis advisors were Harlow Shapley and Arthur Eddington [4].

Schwarzschild and Allan Sandage in the early 1950s. All of the early computational studies concerning stellar structure and evolution were necessarily based on mathematical approximations to the differential equations and power-law expressions for physical variables. This approach resulted in either the generation of analytical expressions or tabulated datasets evaluated with log tables and mechanical calculators [5]. The first fully automated electronic computer models of stars were not to be generated until the mid-1960s, and prominent (among many) in this new field of fast numerical experimentation were Icko Iben (University of Michigan), Rudolf Kippenhahn (Göttingen), Chushiro Hayashi (Kyoto University) and Peter Eggleton (University of Cambridge).

With the ability to perform rapid, brute-force calculations it became possible to include detailed descriptions of such fiendishly complex variables as those relating to the equation of state, energy generation, and the opacity of gases over large ranges of temperature, density, and composition.

All these deep technical matters aside, many of which are still the topic of present-day research, the theory of stellar structure and evolution, by the late 1960s, could be accounted as one of the great triumphs of astrophysics and human ingenuity. Indeed, it is vitally important to understand the measure and workings of stars, since it is largely through the observation of stars and star systems that we learn about the structure of the universe. For all this, however, one of the great and continued issues with respect to the stars is exactly how and why they form.

This being said, the broad brush concepts are clear (and will be described below) and have been known for a century or more. The devil, of course, is in the details, and what is presently not available is a fully comprehensive model that can follow the birth of a young stellar object (YSO) through to the formation of a protostellar core, and on to the appearance of a *bona fide* main sequence star. The task, both theoretically and mathematically, is enormous in such simulations, since the scale changes are truly staggering. In moving from a molecular cloud core to a main sequence star the variations in size, density, and temperature span 6, 19, and 3 orders of magnitude, respectively. Likewise, the physics of gas dynamics, disk accretion, bipolar mass loss, rotation, magnetic field coupling, ionization effects, radiation transfer, shock-front formation, and dust grain evaporation have to be included in the analysis. Additionally, during the formation of a star, some processes run extremely rapidly, while others operate on very long timescales. Even with modern-day supercomputers the study of star formation is a challenge, but fortunately, it is a challenge that many have risen to.

Making Stars the Lockyer Way

Joseph Norman Lockyer was a one of the early pioneers of stellar spectroscopy and stellar classification. Indeed, along with French astronomer Pierre Janssen, Lockyer is credited with discovering helium,[4] the second most abundant element in the universe. In 1869

---

[4] Indeed, the emission lines of helium were observed in the spectrum of the Sun's corona before chemists were able to isolate it from the mineral cleveite (in which helium builds up through uranium decay) and thereafter study it in the laboratory.

Lockyer was the founding and then long-time editor of the prestigious journal *Nature*, and he is additionally credited with being the first ever Professor of Astrophysics—taking on this title and position at the Royal College of Science (now part of Imperial College, London) in 1885. Additionally, he offered one of the first grand synthesis models to describe the formation and evolution of stars.

Lockyer's "temperature arch", or Λ-sequence as we will call it here, was based on extensive spectroscopic observations of elements within the laboratory and of multiple astronomical phenomena, including meteorites, meteors, comets, stars, and nebulae. The Λ-sequence of stellar evolution grew out of his wider thesis of "The Meteoritic Hypothesis" (published in 1890). Lockyer posited that interstellar space was full of extensive meteoroid swarms, and he argued that according to the amount of material in a swarm and how it interacted with others swarms, all the observed astronomical phenomena, from nebulae to stars to novae and comets, could be explained. Indeed, Lockyer stated in "The Meteoritic Hypothesis" that, "All self-luminous bodies in celestial space are composed of either swarms of meteorites or of masses of meteoric vapor produced by heat."

In particular, Lockyer argued that nebulae (all forms) were large swarms of meteorites, and the precursors of stars. As a nebular swarm condensed, Lockyer argued, so it heated up as a result of collisions between the swarm components. This ultimately resulted in the vaporization of the swarm material and the appearance of a high temperature, condensed star. The star, while formed hot, had no envisioned internal energy source, and so, according to Lockyer, it would gradually cool off and eventually liquefy. The final end product, given enough time, would be a cold, dark, liquid body wandering aimlessly through space. In fact, this is the historical reason for the odd sequencing of the temperature axis in the HR diagram.

In Locyker's day the hot spectral-type O and B stars were thought to be the youngest most recently formed stars, while the cool K and M spectral-type stars were thought to be old, liquefied, and cooled-off stars. Under this scheme stars moved down the main sequence as they aged, accordingly, by setting up the x-axis of the HR diagram to run from spectral-type O through to spectral-type M (which is a decreasing temperature sequence). One is also, in essence, plotting an increasing time or age sequence.

This supposed increasing age aspect of the HR diagram, while long-ago abandoned, remains as a fossil within the language of astronomy, with O and B spectral-type stars still to this day being called early type stars, and M spectral-type stars still being called old type stars. The Λ-sequence of stellar evolution described by Lockyer is based on the state of condensation and temperature variation of the star-producing swarm. The star-to-be starts off as a low temperature, low density extended swarm. As it condenses, however, it becomes more compact and hotter (This is the rising part on the left of the Λ symbol). The top of the Λ arch corresponds to the time of maximum heating and brightness, while the descending part of the Λ arch corresponds to the cooling off and liquefaction phase.

Lockyer's intercepting and condensing swarm sequence offered an all encompassing, highly detailed description of star formation, but, of course, as we now know, it is entirely wrong! The point, however, is that it provided a correct mental framework and setting for the idea that star formation is an ongoing, evolutionary process and that stars form in regions where interstellar material is clumped together—that is, in nebula regions. Additionally it set the mental framework through which it became clear that stars are evolving structures: they are born, and having been born they must eventually die. In their death they will be very different objects to that witnessed in their youth.

The Jeans Criterion

The making of stars within a molecular cloud core begins with a victory for gravity followed by a pairing down, fragmentation process. The first step is described by what is known as the Virial theorem, which looks at where the energy resides within a cloud of gas.

The theory of gravitational collapse was first worked out in a detailed mathematical paper by (Sir) James Jeans published in the early twentieth century. Indeed, writing in the *Philosophical Transactions of the Royal Society* for November 1902, Jeans considered the topic of "The Stability of Spherical Nebulae." Jeans analysis began by reviewing an earlier publication by the renowned mathematician (Sir) George Darwin (second son of Charles Darwin), published in 1888, concerning the conditions associated with the stability and collapse of nebulae composed of meteoritic particles. Indeed, it was Darwin who developed a detailed theory for the characteristic shape that would be adopted by rotating, incompressible liquid stars.[5]

The key point in Jeans' (very lengthy and highly technical) analysis, however, was that according to the density and temperature of a specific cloud of meteoritic particles (or a gas) of radius R, there exists a specific scale length R_J, the so-called Jeans length, such that gravitational collapse must occur if $R < R_J$. In general terms, it can be stated that collapse is going to occur if the following energy condition holds true: $|E_{gravity}| > E_{thermal} + E_{magnetic} + E_{turbulence} + E_{rotation}$, where the left-hand term of the inequality is the gravitational potential energy. The terms on the right-hand side of the inequality account for all those mechanisms that act to work against gravitational collapse. The terms listed include the thermal energy of the gas, the support of the gas via an intervening magnetic field, the support of the gas cloud through turbulent motion, and the support of the cloud through rotation. In his original analysis Jeans ignored the effects of magnetic fields, turbulence, and rotation, leaving only the thermal energy term [6]. For typical molecular cloud core values where the temperature is about 10 K the typical Jeans length is of order 10^{14} m, or about 500 AU, and the Jeans mass will be $M_J = 0.05 \, M_\odot$.

Although the Jeans criterion describes a very general collapse condition it does not provide a simple analytic solution for collapse when all possible support mechanisms, rotation, magnetic fields, and turbulent motion are included. It is presently thought, although astronomers are far from achieving universal consensus, that the key mechanisms acting against gravitational collapse are actually turbulence and magnetic fields, the other support mechanisms affording only superficial effects. If nothing else, the Jeans length tells us that once a molecular cloud has started to form a dense core and

---

[5] The idea that stars, and specifically the Sun, might be liquid bodies was largely based on the enumeration of their density. For the Sun the average density is 1400 kg/m³, which, in fact, is about the same as the density as liquid honey. Since most liquids are incompressible, which essentially means that their density is constant throughout, it seemed quite reasonable to astronomers in the late nineteenth century that stars were composed of some hot, incompressible liquid. Additionally, at that time, the continuum part of stellar spectra could, via Kirchoff's laws (recall Chap. 2), reasonably be interpreted as that expected from a hot solid or hot liquid body.

gravitational collapse has begun, then fragmentation into even smaller mass units will most likely occur. This is simply a result of the fact that as a large gas cloud contracts, so its density increases and accordingly the radius for collapse condition becomes smaller and regions contained within the initially large collapsing cloud will find themselves unstable to collapse, and so on.

In this manner the collapse of a molecular cloud core is accompanied by continued fragmentation into smaller and smaller components, and it is the distribution of the final fragmented masses that will determine the mass distribution of the stars formed. Astronomers are still not agreed on exactly how the fragmentation of a collapsing molecular cloud proceeds in detail, but it is known that the mass distribution of newly formed stars, the so-called initial mass function, IMF, is remarkably constant from one star-forming region to another within the galaxy. Indeed, survey work indicates that most stars form within a mass range of 0.3–0.5 $M_\odot$, with the number of higher mass stars dropping of rapidly and decaying as a power law with $\xi(M) = c\, M^{-\alpha}$, where α and c are constant, and where $\xi(M)$ is a measure of the number of stars formed per unit mass interval. In short, however, it appears that star formation favors the production of low mass stars over high mass stars, and this is reflected in the observed numbers of main sequence O and M spectral-type stars—the former stars being much more massive and much less common than the latter low mass stars. Indeed, there are something like 3.8 million stars located within a distance of 230 pc of the Sun, and of these, 2.4 million are low mass spectral-type M stars, but just 1 is a high mass, O spectral-type star. In this same volume of space there will be of order 300,000 Sun-like stars.

Edwin Salpeter (Cornell University), in the mid-1950s, found in the course of his pioneering studies of stars in the solar neighborhood that the mass spectrum constant is $\alpha \approx 2.35$ for $M > 1$ $M_\odot$, and this same constant appears to apply in other star-forming regions within the Milky Way as well as star-formation regions within other galaxies. The number of objects that form with masses less than about 0.3 $M_\odot$ also decreases rapidly, although the behavior of the IMF for masses much below 0.1 $M_\odot$ is highly uncertain.

The basic profile and peak at about 0.5 $M_\odot$ of the IMF appears to be fairly universal, but the cutoffs relating to the upper and lower mass limits for actual star formation vary from one star-forming region to another. Intriguingly, a recent deep study of the Orion Nebula by Holger Drass (Ruhr-Universität Bochum, Germany) and coworkers, using the HAWK-1 infrared camera on ESO's Very Large Telescope (VLT), has revealed that it supports a bimodal IMF. The study finds that the Orion Nebula contains many more low mass objects (brown dwarfs and planetary-type objects) than would otherwise be expected from the standard IMF, a second peak being found at a mass corresponding to 0.025 $M_\odot$.

Figure 5.4 shows four of the newly discovered star formation regions in which low-mass objects (masses > 0.005 $M_\odot \approx 5 \times$ mass of Jupiter) have been found within the Orion Nebula. Two possible explanations for the greater than expected number of sub-stellar objects have been offered. One scenario envisions the formation of such objects in the circumstellar disks around protostars that have chanced to form within small, compact groupings, with subsequently star-on-star gravitational encounters ejecting the smaller objects into the surrounding nebula. A second scenario suggests that the low-mass objects formed directly through the gravitational collapse of material clumps ejected from fragmenting circumstellar disks. Probably both mechanisms are at play.

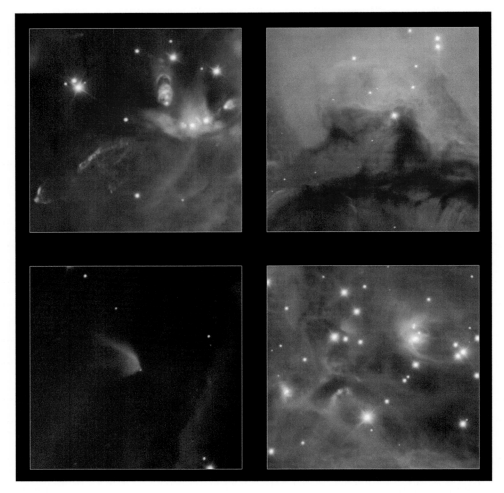

Fig. 5.4 Star-forming regions in the Orion Nebula as observed with the infrared HAWK-1 camera. These images are part of the deepest (that is, faintest) view ever taken of the nebula, and they indicate the presence of more, many more, sub-stellar brown dwarf and planetary-mass objects than would ordinarily be expected. Image courtesy of ESO

Young Stellar Objects

Star formation is all about the steady march towards higher and higher density. The driving force is gravity, and the end result, if nothing opposes it, is death by black hole. Indeed, a star will literally shrink itself out of existence if gravity wins out over competing forces. Figure 5.5 is an attempt to illustrate the indomitable motion towards smaller size and higher density as a molecular cloud core evolves through its star phase to that of a stable end state.

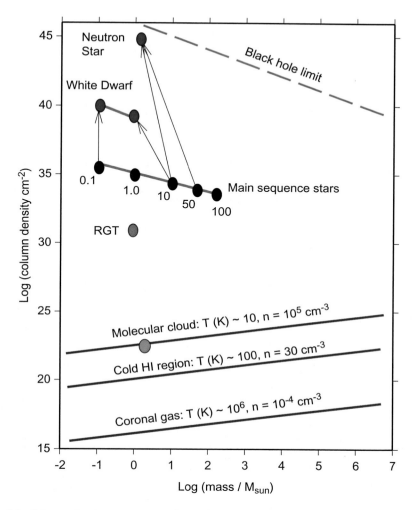

Fig. 5.5 Column density versus mass for various phases of the interstellar medium, main sequence stars, giant stars, and degenerate end-phase objects. The characteristic values for the column density and mass of the Bok globule Barnard 68 are shown in the diagram (*green circle*), and these provide an example of those conditions applicable to a dense molecular cloud core likely to begin producing stars in the relatively near future

The diagram actually shows column density verses configuration mass, the idea being that the column density, n, is an expression that relates to the bulk density of material and the physical size of the region within which the matter is contained.[6]

---

[6] Formally, the column density is the product: $n = 1 \text{ m}^2 \times (\rho / m_H) \times 2R$, where R is the radius of the assumed spherical structure, ρ is the density, and m_H is the mass of the hydrogen atom. The factor of 2 enters in since it is the diameter of the spherical region that is required in the calculation of n.

In the lower part of the diagram in Fig. 5.5 the column density for various mass coronal, cold HI, and molecular gas clouds are shown. The temperature label on each of these three curves is not actually used to calculate the column density but is rather given for reference. And again we make the simplifying approximation that the clouds are spherical structures. The evolutionary path of a given mass protostar is launched from the line corresponding to molecular clouds, and assuming no mass loss, its life path is essentially that of a vertical line moving upwards in the diagram. Once the gravitational collapse of a molecular cloud core has begun, its first stop, and indeed its longest stop, in the journey towards a final resting place, is that of the main sequence. Here it remains, converting hydrogen into helium within its central core, gravity being held in check via the pressure gradient established throughout the stars interior.

With the exhaustion of hydrogen in the central core, a star will reverse its path briefly, moving downwards in the diagram as it expands to become a bloated red giant star, converting helium into carbon within its central core. The location of the Sun when it is at the red giant tip (RGT) just prior to the onset of helium burning is shown in the diagram. At this time it will be some 160 times larger than at present. What happens after the red giant phase depends upon the initial mass.

Stars between 0.1 and 8 times the mass of the Sun end their days as diminutive but stable white dwarfs. The Sun will eventually evolve into a white dwarf (of mass ~0.6 $M_\odot$) in about 6 billion years from now. White dwarfs, if single and left alone, will cool off over billions of years to become frigid, zero-luminosity black dwarfs, and in this case the march of gravity is brought to a complete and indefinite halt. Stars more than 8 times the mass of the Sun end their days as neutron stars. These objects are even smaller and more tightly packed than white dwarfs, but once again, if left alone and single, they will remain stable for all time, and gravity once again loses the contraction battle.

Up to about 50 times the mass of the Sun it is generally agreed that after the supernova end stage what is left of the core of a star will form a neutron star (with a maximum stable mass of about 3 times that of the Sun). Above an initial mass of about 50 $M_\odot$, the end state for the core of a star after its supernova disruption is that of a black hole. In many ways it does not make much sense to talk about the column density of a black hole, since all the matter is crushed into a singularity at the center, with the density there becoming infinite. However, the line labeled 'Black hole limit' in Fig. 5.5 corresponds to the number density versus mass with the simplifying assumption that the radius is the so-called Schwarzschild radius (also known as the event horizon radius) and that the density is simply the mass of the black hole divided by its volume.

The story of star formation begins once the fragmentation and Jeans criterion sorting of the initial collapsing molecular cloud core comes to an end. The idealized picture now is one of a uniform temperature, spherical cloud with a density profile that increases towards its center as the inverse radius squared: $\rho(r) \sim r^{-2}$. This density profile actually applies to what are called isothermal spheres. Such idealized (technically infinite radius) spheres have the same temperature everywhere and are hovering on the brink of gravitational instability. So-called Bonner-Ebert spheres correspond to finite-sized isothermal spheres that are constrained by some external pressure. Although the mathematical details of such structures need not concern us here, the essential idea being adopted is that the collapse of the cloud, to produce a specific star, proceeds from a state that is close to that of an isothermal sphere, with the equations describing the actual time variation of the temperature, density, and mass distribution as a function of the radius necessarily being solved for numerically on a computer.

With an initial state similar to that of an isothermal sphere, each radial shell of the cloud will begin to collapse at a free-fall rate. Given that the collapse time varies as the inverse square root of the density, $t_{coll} \sim \rho^{-\frac{1}{2}}$ [1], and that for an initially isothermal configuration the density varies as the inverse radius squared, $\rho(r) \sim r^{-2}$, so the collapse timescale varies directly with the radius: $t_{coll} \sim r$. This result tells us that the collapse of the cloud proceeds from the inside out. Accordingly, a small central core builds up rapidly and then accretes material over an extended period of time. Remarkably, however, it turns out that to a good approximation the accretion rate at the cloud center is constant over time [7].

The accretion rate at the center of a collapsing cloud is essentially constant during the star building process, but the way in which the material from the outer regions of the cloud approaches the core is not that of simple radial collapse; rather, the infalling material forms an accretion disk around the centrally forming protostellar core. This disk forms due to the fact that the collapsing cloud itself is rotating, albeit very slowly, and because of a conservation law that applies to the angular momentum of the infalling material. The upshot of all this [8] is that the infalling material forms a disk-like structure around the central protostellar core, with the inner regions of the disk spinning more rapidly than the outermost regions (Fig. 5.6, and recall Fig. 2.14). Rather than the material from the cloud falling

Fig. 5.6 Artist's impression of a dusty accretion disk surrounding a newly formed protostar. Material from the collapsing molecular cloud falls onto the disk and then migrates towards the central region, where it is accreted by the proto-stellar core. When magnetic fields are included in the cloud collapse simulations, jets of outflowing material can form along the spin axis of the system. Image courtesy of ESO/L. Calçada/M. Kommesser

directly onto the proto-stellar core, therefore, it falls onto an extended disk and thereafter spirals inward towards the central regions, where it is eventually accreted by the core.

Accretion disks work, that is, allow material to move inwards, in spite of the apparent restriction of angular momentum conservation [8], because of gas viscosity. If we imagine the disk to be made up from a whole series of concentric circular rings, with each inner ring rotating more rapidly than the next outermost one, then the outer edge of each ring "rubs" against the innermost edge of its surrounding ring. This contact rubbing (determined by the viscosity) between each ring pair results in a torque being applied to the outer ring by the inner ring. In effect the inner ring, which is rotating more rapidly, is continually trying to speed up the rotation rate of the outer ring. This torque amounts to an outward transport of angular momentum through the disk.

To compensate for this outward transport of angular momentum, due to the viscosity, material within the disk moves inwards, carrying its higher angular momentum with it. In the end, it is a balancing act that comes about. Angular momentum is conserved even though material is spiraling inward through the disk to be accreted onto the central proto-stellar core. In addition to allowing material to flow through the disk, the viscosity between each ring in the disk results in it becoming heated, with the temperature in the innermost regions being higher than that in the outer regions. Indeed, it is the characteristics of the energy radiated into space by the heated disk (and eventually by the growing proto-star) that is used to define the various categories, as introduced by Charles Lada (Harvard University) in 1987, of young stellar objects (YSOs).

Figure 5.7 is a schematic diagram showing the energy spectrum characteristics (recall Fig. 2.14) ascribed to YSO Classes I, II, and III, along with a sketch of the various system configurations. The classification is based on the slope of the spectral energy distribution in the wavelength region between 2.2 and 20 μm, and follows the evolution of the system from that of early cloud collapse to the formation of a proto-stellar core surrounded by an actively accreting disk (Class I system). For the Class I systems it is the infrared radiation of the disk that dominates the energy spectrum, with a peak intensity being seen in the far infrared at a wavelength of $\lambda \sim 100$ μm.

The protostar at this stage accounts for a relatively small amount of the emission. As the system evolves from a Class I to a Class II designation, it is the emission from the proto-stellar core that begins to dominate, and the peak intensity shifts towards a shorter wavelength of $\lambda \sim 1$ μm. As the accretion onto the disk falters, so the disk becomes increasingly less luminous and cooler, and ultimately, at Class III, it is only detectable as a slight infrared excess superimposed on the now dominant spectrum of the protostar.

The star formation rate (SFR) in any particular molecular cloud can be gauged upon the observed number of YSOs and the characteristic lifetime of the YSO phase. Specifically, the SFR describes the rate at which interstellar material is being converted into stars, and it is expressed in units of solar masses per year. Accordingly, when a molecular cloud can be resolved into active regions, and the individual YSOs can be counted, so the star formation rate is given as $SFR = N_{YSO} \, M_T / \tau$, where M_T is the typical mass of a newly formed star—which, from the characteristics of the IMF, is of order 0.5 M$_\odot$ and where $\tau \approx 2$ million years is the characteristic lifetime of a YSO.

In those situations where the star-forming cloud cannot be resolved into individual YSOs an integrated luminosity can be used as a proxy measure. This technique once

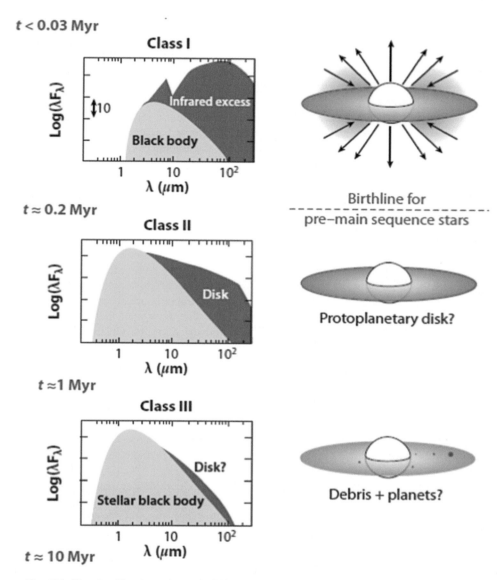

Fig. 5.7 The classification scheme for YSOs. To the left are a series of representative spectra for the Class I, II, and III YSOs. The contribution to the spectrum from the disk is shown in *red*, while the contribution from the protostar is shown in *yellow*. The class sequence follows the evolution of the system towards that of a young star approaching the main sequence, surrounded by a quiescent disk in which planets might eventually form. Image courtesy of Swinburne Astronomy Online

calibrated within our own Milky Way is particularly important when looking at distant galaxies. Many measures for the SFR are available, and they span the spectrum range from the X-ray to the UV, infrared and radio. The UV measure of the SFR is based on the formation of massive O and B spectral-type stars and their associated HII clouds. The high surface temperatures of these stars results in the copious emission of UV photons and in the formation of a strong hydrogen-alpha (Hα) emission line. Accordingly, the integrated Hα luminosity of a molecular cloud or distant galaxy can be used, after calibration, to determine the SFR. In contrast, the infrared SFR measure is based on measuring the interstellar dust grain temperature, which will vary according to the amount of heat being provided by newly formed, hot stars.

With reference to Fig. 4.11, the deduced SFR in the Orion molecular cloud are 10^{-3} $M_\odot$/year for Orion A and 10^{-4} $M_\odot$/year for Orion B. These star formation rates are in fact typical of most star-forming regions in the solar neighborhood. Indeed, with reference to Fig. 4.17, the deduced SFRs for the Perseus, Ophiuchus, and Taurus (TMC-1) molecular clouds are 10^{-4} $M_\odot$/year, while that for the RCr A cloud is somewhat smaller at 3×10^{-5} $M_\odot$/year. Extending the observations to the entire Milky Way Galaxy it is estimated that the global SFR is of order 1–3 $M_\odot$/year. A similar rate is deduced for the Andromeda Galaxy (M31), with the SFR being ~0.5–1.0 $M_\odot$/year. Like the waistlines of astronomers, however, galaxies come in all manner of shapes and sizes, and some galaxies are observed to be generating more stars per unit time than their neighbors. The Whirlpool Galaxy (M51), for example, has an SFR of some 5–7 $M_\odot$/year, while the Starburst Galaxy (M82) has a measured SFR of several hundred solar masses per year.

A remarkable empirical relationship between the SFR per unit area (Σ_{SFT} in units of $M_\odot$/year/kpc^2) and surface gas density (Σ_{gas} in units of $M_\odot$/pc^2) in our galaxy was discovered by Maarten Schmidt (CALTEC) in 1959, and later extended to other galaxies by Robert Kennicutt (then at the University of Arizona) in 1998. This so-called Schmidt-Kennicutt law reveals that a very close relationship exists between the surface gas density and the rate at which stars are being formed, with the SFR being higher in those regions were there is more gas and dust. Specifically, the observations indicate that $\Sigma_{SFT} \sim (\Sigma_{gas})^n$, with the data indicating that $n \approx 1.4$.

What is equally as remarkable as the existence as the Schmidt-Kennicutt law is that the efficiency of star formation, irrespective of the SFR, is invariably found to be no more than one or two percent. It would appear that in essentially every situation whenever and whenever stars begin to form, numerous mechanisms (ionizing winds, bi-polar jets, supernova explosions) conspire to make sure that only a small fraction of the potential star-building material is actually converted into stars at any one time.

The Stellar Birth Line

The inside out collapse of a star-forming cloud results, as indicated earlier, in the formation of a small central core, and this core grows in mass via disk accretion. The manner in which material falls onto the proto-stellar core, however, is far from a gentle or particularly well understood process. Indeed, the material falling onto the proto-stellar core does so at supersonic speed and accordingly creates a hot accretion shock at the core surface. The luminosity at the accretion shock is determined by the accretion rate and the amount of gravitational energy released by the material falling in from the surrounding cloud. A small fraction of the shock energy, however, also goes into heating the outer layers of the proto-stellar core, and the temperature at the surface of the core is soon driven to a level well in excess of several thousand degrees. This high surface temperature sets up a zone around the core where the dust within the accretion flow will be vaporized and where atoms and molecules will become ionized.

It is the ionization of the inflowing material that proves critical with respect to the formation of jets. Detailed numerical simulations show that in the presence of magnetic fields, rotation, and ionization, the magnetic field becomes tightly wound around the spin axis of the system, and by entraining ionized material from the accretion flow re-directs some material outward in the form of two jets, one above and one below the plane of the disk (see Fig. 5.7). These jet features are characteristically associated with newly formed T Tauri stars and Herbig-Haro Objects (HHOs).

Named after George Herbig and Guillermo Haro, who first studied and cataloged them in the 1950s, HHOs are small knots of optical emission, invariably found in star-forming regions, that are typically moving away from young embedded stars at speeds of several hundreds of kilometers per second (Fig. 5.8). The emission features associated with HHOs are caused by the shock heating of those regions where the fast-moving material in the jet encounters the more slowly moving material ahead of it.

We now have a basic picture for the formation of a protostar. The molecular cloud core region collapses from the inside out, first forming a small proto-stellar core. Due to the small initial cloud rotation an accretion disk forms around the core, and the core mass increases at a steady rate, becoming more and more massive and more and more luminous as time goes by. Given a typical accretion rate of 10^{-6} $M_\odot$/year, it accordingly takes of order 1 million years to produce a one solar mass star. Heating and ionization of the material at the inner edge of the accretion disk is picked up by an entrained magnetic field, and bi-polar jets are formed along the spin axis of the star.

Eventually all of the material that is going to be accreted from the natal molecular cloud finds its way into the disk, and the final stages of proto-stellar evolution can now proceed. Indeed, at this stage the remnant accretion disk may have already started the planet-building process and the protostar is observable as a low mass ($M < 2$ $M_\odot$) T Tauri pre-main sequence star, or as a more massive Herbig Ae/Be pre-main sequence star. These latter stars were first described by George Herbig in the 1960s, with the A and B indicating the characteristic spectral type of such objects, and the e indicates the presence of anomalous emission line features.

During the early formation and accretion phase the central proto-stellar core transports energy throughout its interior by convection, this being the literally turning over and churning of its internal gas. In a famous research paper published in 1962 Chushiro Hayashi showed that such fully convective low-mass cores could not be dynamically stable unless their surface temperatures were higher than a critical value of about 4000 K. Accordingly, it transpires that a young proto-stellar core is born with a relatively high luminosity (controlled by the accretion rate), a large radius, and a surface temperature of order 4000 K.

Fig. 5.8 Herbig-Haro Objects 901 and 902 in the Carina star formation nebula. In each case the newly formed stars are still enshrouded in their natal dust columns, but their associated bi-polar jets have pushed outward and produced distinctive bow shocks (the actual HHOs) in the surrounding interstellar medium. Image courtesy of NASA/HST

Additionally, subsequent evolution sees it descend to the so-called Hayashi boundary. The surface temperature remains approximately constant at this time, while the luminosity and radius become increasingly smaller (see Fig. 5.9). During this phase of contraction the central temperature of a star steadily increases, and eventually the mode of energy transport within its interior changes. Indeed, it changes from being fully convective to being fully radiative.

In this latter situation the outward transport of energy is via photons rather than the bulk motion of the internal gas. In this fully radiative state the protostar continues to contract, but now its surface temperature and luminosity begin to steadily increase. The protostar is now moving towards the acquisition of stardom and the attainment of its mass-appropriate main sequence location in the HR diagram.

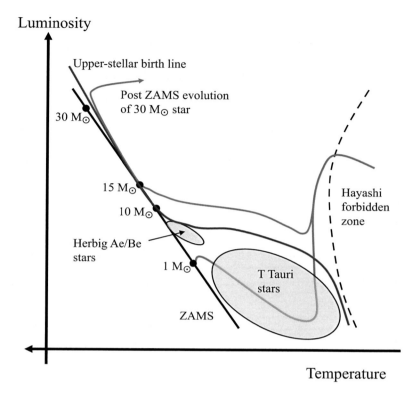

Fig. 5.9 The stellar birth line. In this schematic HR diagram the long diagonal line corresponds to the zero-age main sequence (ZAMS). This is the locus along which newly formed, uniform composition stars, just initiating hydrogen fusion reactions, will fall. To the right is the Hayashi forbidden zone, in which no dynamically stable solutions for fully convective, low-mass cores can be obtained. The stellar birth line (*blue*) indicates where newly formed pre-main sequence stars first appear as either T Tauri-like stars, Herbig Ae/Be stars, or luminous O stars. Schematic evolutionary tracks for a solar mass star and a 30 $M_\odot$ star are shown in *red*. The accretion rate for the 30 $M_\odot$ star is assumed to be 10^{-5} $M_\odot$/year, and it only becomes optically visible after emerging from the upper-stellar birth line in a post zero-age main sequence state of evolution

The so-called radiative or Henyey tracks of protostars, as they approach the main sequence, were first studied in detail by Louis Henyey and coworkers in the mid-1950s. Eventually, the temperature at the center of the protostar becomes hot enough for nuclear fusion to begin, and it is at this time that the star stops contracting and settles into its long-lived, stable main sequence phase (see later).

With the onset of steady nuclear burning within the central regions the manner in which energy is transported inside of a star changes once again. For stars less massive than about 0.4 $M_\odot$ the energy transport is almost entirely by convection. Between 0.5 and 1.5 $M_\odot$ stars develop radiative cores that are surrounded by convective envelopes, while beyond

1.5 $M_\odot$ the stars develop increasingly large convective cores with concomitant smaller radiative envelops. It is the progression in initial mass and the changing modes by which internal energy is transported that determine the properties of a star on the main sequence.

All of the early accretion and the proto-stellar core's evolution along the Hayashi boundary occur behind an optically shrouded curtain of gas and dust. Once the accretion has largely stopped, however, the protostar begins to emerge from its natal cocoon and becomes visible at optical wavelengths. The boundary in the HR diagram where a proto-star first emerges from its dust and gas shroud is called the stellar birth line (see Fig. 5.9). Where the star actually crosses the birth line in the HR diagram depends on its mass and the reserve of material that it has yet to accrete from its surroundings. Low mass stars, with $M < 2 \, M_\odot$, will be first observed as T Tauri stars, while more massive stars, with $2 < M/M_\odot < 10$ will appear closer to the main sequence and be identified as Herbig Ae/Be stars.

The idea of the stellar birth line, where stars are first observed at optical wavelengths, was first discussed by Stephen Stahler (Harvard University) in 1983. More massive stars may not be visible optically until they are in an advanced hydrogen burning phase, the star still accreting material even though it has initiated fusion reactions within its core. These more massive stars are thought to power what are known as ultra-compact H II regions, and they will eventually become visible at a location beyond the upper-stellar birth line in the HR diagram (see Fig. 5.9).

This author's doctoral thesis (University of Western Ontario, 1993) was concerned with the formation of 30 $M_\odot$ stars, and a numerical computer code was used to follow the growth of such stars from an initial 1 $M_\odot$ fully convective core, located close to the Hayashi boundary, all the way up and along the main sequence, and eventually through to emergence at the upper-stellar birth line. Indeed, to produce a 30 $M_\odot$ star, at an accretion rate of $10^{-5} \, M_\odot$/year, takes of order 3 million years, and this is a significant fraction of the star's overall main sequence lifetime. Under the accretion scenario the last half of the star's eventual mass, accounting for the final 15 solar masses, was accreted after the onset of nuclear fusion reactions in the core. To this day there is still continued debate about the actual formation circumstances of the most massive stars, and it appears most likely that for masses much above 20–30 $M_\odot$ the formation process is driven by cloud–cloud collisions or by rapid cloud compression (by a nearby supernova explosion). In this revised situation the star-making material is brought together much more rapidly and at a much higher accretion rate than would otherwise be the case.

Acquiring Stardom

Gravity, like a young child, knows no boundaries, and if given the chance it will crush matter out of all existence. Although the weakest of the fundamental forces, gravity operates on all scales, from the very large to the very small, and its memory is very long. Indeed, the life of a star is one long struggle of holding gravity at bay, and while sometimes the star wins the long-term tussle, at other times it loses. Stars may well be long-lived and apparently stable structures, but gravity is always there in the background—it never sleeps.

Stars cannot exist without gravity, but if they drop their guard for just a few hours then gravity will collapse them down to a singularity. As with most physical contests, the end result lies in the balance between those that are pushing and those that are pulling—a dynamic truce, as it were. For a star, gravity is working to pull the material body into as small a volume of space as it can, but to counterbalance this, the star utilizes its internal heat to generate an outward-pushing pressure gradient. The pressure gradient literally holds gravity in check and supports the star against collapse, with each layer inside of the star supporting the weight of all the other layers above it. A quivering stalemate between push and pull is achieved, and the star hovers breathlessly in a state of dynamical balance, with each proponent, gravity, and the pressure gradient, holding their ground but continually testing for any weakness in their opponent.

If gravity wins some ground, the star contracts, but in response to this its interior heats up, and this produces a steeper outward pressure gradient, and the ground won by gravity is taken back. If the pressure gradient gains ground, then the star expands, but now its interior cools slightly, and this reduces the pressure gradient and gravity wins back its lost ground. In a continual game of give and take-back a star's radius remains constant, in hydrostatic equilibrium, over long intervals of time.

The dynamic, or collapse timescale τ_{dyn} for a star, on the assumption that all outward pressure forces, in a miraculous manner, suddenly disappear is calculated in the same way as that for a molecular cloud. In this manner, recall $\tau_{dyn} \sim \rho^{-1/2}$, where ρ is the average density of the star [1]. From this relationship we see that the higher the average density of a star, the more rapid it will collapse if gravity wins out over all other resistive forces. For the Sun the average density is 1400 kg/m^3 and the collapse time is about 30 minutes, and this tells us how finely tuned the balance condition must be. If, for example, gravity is imagined to be very slowly winning the battle against the pressure gradient by a factor of ε times, the gravitational force term, then the fact that the Sun is already 4.5 billion years old indicates that $\varepsilon < 10^{-28}$ [1]. On this basis to simply say that the Sun is highly stable seems like a gross understatement.

A star can remain in hydrostatic equilibrium provided it can keep its interior hot, and this requires that it generates energy within its interior. A star must generate energy within its interior since it would other wise cool-off as a result of energy loss at its surface. The fact that stars shine (and therefore have a measurable luminosity) means that they must ultimately die, since no fuel supply can last for ever.

The question concerning how long stars might shine was one of the first great discussions of nineteenth century astrophysics. The Sun's measured energy output per unit time, its luminosity $L_\odot$, is known to be 3.85×10^{26} Joules per second. Additionally, the Sun's mass $M_\odot$, is known to be 2×10^{30} kg. Measure for measure, therefore, the average amount of energy the Sun needs to produce per kilogram of its mass is 0.0002 Joules per second. This is a ridiculously small number. The average power output of 15 selected cyclists who rode the 2005 Tour de France, which covered 3592.5 km over 21 days of racing, in 86 hour 15 minutes by the winning cyclist, was found to be 3.2 Joules per second per kg [9]. That's some 1600 times more power generated per kg than the Sun. What we need to remember at this stage, however, is that a Grand Tour cyclist has only to maintain a high average energy output per kg for some 5 hour per day on each of the 21 days of racing (which, of course, is an incredible human feat). The Sun, in contrast, has to maintain its energy output continuously for billions of years, and it never gets to eat or sleep.

Having identified longevity as the key issue with respect to stardom—that is, without an internal energy source a star will not remain stable for long—what therefore does a star "burn" in its interior to maintain the necessary pressure gradient to keep it in hydrostatic equilibrium? To answer this, the term burn should first be qualified.

When something, wood or coal, say, burns, it is a chemical reaction that is taking place, and this is a catalytic reaction involving an initial heat source, to get the process started, and then a supply of oxygen is need to sustain the reactions once begun. The amount of energy released in burning reactions is then dependent on what it is that is actually burning. In burning 1 kg of coal, for example, 25 mega-joule's worth of energy is given off.

The next issue relates to the length of time T_{burn} over which the burning can be sustained. This is a straightforward calculation and simply relates to the amount of energy available in the fuel supply E divided by the rate at which the fuel is being converted into energy L. Accordingly: $T_{burn} = E/L$. If, as a grand Victorian industrialist might well have speculated, the Sun is made of coal, then $T_{burn} = 25 \times 10^6 (M/L)_\odot \approx 4000$ years. To a highly devout early Victorian industrialist this timescale for a coal-powered Sun would have been perfectly fine; the universe, after all, was only a few thousand years old. Unfortunately for this argument, however, the mid-Victorian geologists, evolutionary biologists and eventually physicists [10] realized that Earth was much older than 4000 years, and accordingly a longer-lasting energy source was required to power the Sun.

Moving beyond chemical burning, a glimmer of hope for the astrophysicist was introduced in the late 1890s when Marie and Pierre Currie discovered the radioactive properties of ^{226}Ra (in the form of radium chloride, radium itself eventually being isolated by Marie in 1910). In a 1 kg mass of pure ^{226}Ra, some 37 trillion atomic disintegrations will take place each second, and each of these decays will liberate 4.871 MeV of energy. In total, therefore, the energy available from the decay of 1 kg of pure ^{226}Ra amounts to some 28.5 Joules per second. If the Sun was powered by the radioactive decay of radium it would require a total of 2.26 Earth masses (about 1.35×10^{25} kg) of ^{226}Ra to sustain it at its present luminosity.

This amount of radium, while not wholly unimaginable, might at first glance seem to have solved the problem of powering the Sun, but again it is the lifetime issue that is the downfall. The half-life of ^{226}Ra is just 1600 years, and though one might tweak the number a little, a total fuel supply lifetime of no more than perhaps three half-lifetimes, that is about 5000 years, is going to be available. This lifetime is a little greater than that from a pure coal-powered Sun, but not by much, and certainly not enough to satisfy the biologists and geologists.

Even plutonium, first isolated in 1940, cannot solve the Sun's longevity problems. Some 5.5×10^{28} kg (1/37th the actual mass of the Sun) of pure ^{240}Pu could provide for the Sun's present luminosity, but it might do so for perhaps 2 or 3 half-lifetimes, and this amounts to just 16,000 years. Even if the entire Sun was made of uranium-238, it couldn't generate enough energy per second to provide for the observed luminosity, although, ironically, the half-lifetime of ^{238}U is about right for the present age of the Solar System—4.5 billion years.

A hybrid radioactive decay model for powering stars was developed by British physicist (Sir) James Jeans in the early decades of the twentieth century. In a letter published in the August 2, 1917, issue of the journal *Nature*, Jeans argued for the direct transformation of matter into energy. This scenario was not quite the matter–antimatter annihilation that we are more familiar with today, but involved the direct annihilation of matter with the consequent liberation of energy [11].

One of the criticisms leveled against the Jeans energy generation model was that it implied the Sun must have been heavier in the past, and this would have direct consequences for the stability of planetary orbits. Needless to say, this was perhaps the least of the problems associated with the Jeans model. Although it is now known that radioactive decay is incapable of solving the Sun's energy and longevity crisis, the answer to its problems do lie within the realm of nuclear physics.

Before considering how the Sun and stars usually generate internal energy, two additional possibilities need to be dealt with. One builds on the ideas presented by Joseph Lockyer, and the other brings us back to gravity. While the chemical burning versus radioactive decay debate was raging, two other mechanisms for generating energy were being discussed. One, in principle, could power the Sun forever, and the other offered a lifetime of at least 10 million years. Joseph Lockyer proposed in 1890 that stars were formed through the condensation of vast swarms of meteorites. To this he added that perhaps once formed, a star might continue to be sustained through the energy of infalling meteorites. In this case, the Sun's luminosity is maintained from the mechanical supply of heat in the form of the kinetic energy of impacts. To power the Sun, however, a combined mass of 2×10^{15} kg of material would have to strike the Sun's surface every second, which is equivalent to about 6 million Empire State Building's worth of matter per second. The beauty of Lockyer's accretion idea is that, provided material keeps falling onto the Sun, it can continue to shine for as long as one requires. The critical problem with Lockyer's accretion idea, however, is that no interstellar swarms of meteorites actually exist [12].

The second method of interest, and indeed one of actual necessity, is that relating to gravitational collapse itself. Although hydrostatic equilibrium comes about when the internal energy source of a star generates enough internal heat to enable the pressure gradient to support the weight of overlying layers, prior to this condition coming about a star can gain energy by getting smaller—that is, collapsing slightly and in the process liberating gravitational potential energy.

This idea was developed by Hermann Helmholtz and Lord Kelvin in the later half of the nineteenth century. Here, by a slow and steady contraction, a star can convert its potential gravitational energy into internal heat energy, and to power the Sun at its present luminosity this would require the radius to change by about 10 m per year. While this amount of contraction would be unnoticeable, not even measurable from Earth, on timescales shorter than centuries it would eventually be revealed as a very obvious change in the Sun's angular size. Even if powered by slow gravitational contraction, a star still cannot last forever, since eventually it will become a black hole. The characteristic contraction time for the Sun to radiate away its gravitational potential energy, the so-called Kelvin-Helmholtz timescale, τ_{KH}, is about 30 million years.

This then is the issue. If there is no internal energy source the Sun will gradually collapse under its own gravity and shine through the liberation of gravitational potential energy. Furthermore, if there is nothing to oppose the gravitational collapse, then the Sun will eventually become a black hole with an event horizon radius of about 3 km. The only way to halt the gravitational collapse is to provide the Sun with either an external (*a la* Lockyer accretion) energy source, or via an internal one. It was the great (Sir) Arthur Eddington who first worked out in the mid-1920s how the Sun can save itself from collapse through the workings of an internal nuclear energy source.

Unlike Jeans, who invoked a hybrid, radioactive-like decay mechanism as a means of generating internal energy, Eddington developed his ideas around the direct conversion of matter into energy as had been outlined in Einstein's theory of special relativity, first published in 1905.

The first step in Eddington's scheme was based on an experimental result derived by Francis Aston in the early 1920s. Aston was a chemist working at the Cavendish Laboratory at the University of Cambridge, and he specialized in measuring the minute mass of atomic nuclei. Indeed, Aston pioneered the development of the mass spectrograph and was awarded the Nobel Prize for Chemistry in 1922 for his efforts. What caught Eddington's eye, however, was the result that the mass of four hydrogen atoms was not the same as the mass of one helium atom.[7] It was as if $1+1+1+1$ did not equal 4. This was important, since Eddington reasoned that if the most abundant and simplest atom, hydrogen, was going to be converted into the next simplest and next most abundant atom, helium, at least four hydrogen atoms would be required for the conversion.

The experimental results of Aston implicated, however, that $4H = He - \Delta m$, where the mass deficit determined by Aston's experiments was $\Delta m = 0.007 \times 4m_H$, where m_H is the mass of the hydrogen nucleus (which is just the mass of one proton). Somehow, 0.7 % of the mass of the four hydrogen atoms was lost in the process of producing a helium nucleus. This loss, Eddington argued, was not some mass accounting error or the indication of some new particle being generated, but was the result of mass being converted into energy.

With this understanding in place, Eddington could apply Einstein's mass-energy equivalency formula and argue that in the conversion of four hydrogen atoms into one helium nucleus the amount of energy liberated per conversion would be $\Delta E = 0.007 \times 4m_H c^2 = 4.2 \times 10^{-12}$ J. This liberated energy, Eddington argued, could be used by the stars to keep their interiors hot and thereby maintain the requisite internal pressure gradient to remain stable, and to replenish the energy lost into space at their surface. The Sun has a measured luminosity of 3.85×10^{26} Joules per second, and this indicates that of order $L_\odot / \Delta E \approx 10^{38}$ conversions of $4H \rightarrow He$ must take place per second. Since the number of hydrogen atoms in the Sun is of order $M_\odot / m_H \approx 10^{57}$ only a minute fraction of its hydrogen need be involved in generating its energy at any one instant.

Having established the idea of a new energy generation mechanism, Eddington next estimated how long a star might be powered by fusion reactions. The nuclear timescale is accordingly evaluated as $\tau_{nuc} \approx 0.007\, c^2\, M/L$, where it is assumed at this stage that the star, of mass M and luminosity L, is entirely composed of hydrogen, and that all of the hydrogen in the stars is available to be converted into helium. Using numbers for the Sun, the nuclear timescale evaluates to $\tau_{nuc} \approx 3.3 \times 10^{18}$ s $\approx 10^{11}$ years—a timescale very much longer than that of the Kelvin-Helmholtz timescale of 30 million years.

This now lengthy energy-generating nuclear timescale for the Sun can comfortably accommodate the age set for Earth by the geologists, and the apparent conflict of Earth

---

[7] At the time that Eddington was developing his ideas the structure of atomic nuclei was not well understood, and it was assumed that nuclei were entirely composed of protons. James Chadwick discovered the neutron, which has essentially the same mass as the proton, in 1932, and this provided the present-day picture of atomic nuclei being composed of both neutrons and protons.

being older than the length of time that the Sun can shine for is removed. The age conflict, of course, could be removed only if nature had actually found a way of transforming hydrogen into helium, and exactly what that process might be was then entirely unknown. Writing in his *The Internal Constitution of the Stars* (1927), Eddington captured the core of the problem: "The difficulty is that from the physicist's point of view the temperature of the stars is absurdly low." Given that the Sun has a central temperature of some 15 million degrees Eddington's statement seems utterly remarkable, but it is, in fact, entirely true in terms of classical physics [13]. At issue is how to bring four initially independent protons together so as to make one helium nucleus.

The physics of the process requires that the protons must be brought into extreme close contact, overcoming the repulsive Coulomb barrier due to their electrical charge, and thereby allowing the attractive nuclear binding force (the strong force in modern parlance) to hold them in place. Now, the higher the kinetic energy of the protons, the closer they can approach each other, but it transpires that even at temperatures of many millions of degrees, the protons will still not have enough energy to overcome the requisite Coulomb barrier. It was as if nature had put in place an impenetrable wall against which the protons could not proceed.

Additionally, the probability that four protons might actually be brought close enough together, at the same time, in order to react with each other is (even by astronomical standards) vanishingly small. There is no respite in the problem, however, even if one tries to build up a helium nucleus one proton collision at a time. The problem is that the diproton, or helium 2 nucleus, is highly unstable, and even if one could produce such a pairing of protons, the Coulomb barrier is still stronger than the nuclear force, and the nucleus will disaggregate instantaneously.

In order to solve the problem of stellar energy generation two fundamentally new experimental results had to be established. The first key result was described by George Gamow (then at the University of Copenhagen) in 1928 and was based on the new ideas of quantum mechanics. Specifically it was the notion of quantum tunneling that Gamow introduced. Importantly, for the hydrogen fusion process to work, this tunneling capability allows particles to penetrate the Coulomb barrier even if their kinetic energy would indicate that it cannot be done [13].

In the wonderful quantum mechanical world the apparently impenetrable Coulomb barrier is circumvented by the particle disappearing on one side of the wall and instantaneously re-forming on the other. The second key result that needed to be put in place before hydrogen fusion could be understood was that relating to the weak nuclear interaction, which enables β-decays. Enrico Fermi introduced the detailed theory of β-decay in 1934, and at the same time predicted the existence of a new particle—the neutrino, first detected experimentally in 1956.

During a β-decay a neutron can spontaneously decay into a proton, with the ejection of an electron and a neutrino: $n \rightarrow p + e^- + \nu$. The inverse of this process can also take place, with the proton transforming into a neutron with the ejection of a positron and a neutrino. It is this latter inverse β-decay that allows stars like the Sun to shine—a result first described in detail by Hans Bethe and Charles Critchfield in 1938. Indeed, it is the combination of quantum tunneling and a simultaneous inverse β-decay that allows the first step in the stellar energy generation cycle to proceed.

The picture now developed is that when two protons approach each other at the same time as the quantum tunneling allows them to penetrate the Coulomb barrier one of the protons undergoes an inverse β-decay to produce a neutron, with the formation of a deuterium nucleus: $p+p \rightarrow {}^2H+e^++\nu$. This first step occurs only very rarely, and on average a proton at the center of the Sun will have to wait some 10^{11} years before undergoing such a deuterium-forming event with another proton.

Here the glory of statistics comes to the Sun's salvation, and we also learn why the Sun has to be as massive as it is. Although a proton has to wait *on average* a 100 billion years before it is likely to enable the generation of a deuterium nucleus, many protons will undergo the reaction much sooner, and others will take even longer. Accordingly, if a star has many, many protons within its interior, and recall the Sun has of order 10^{57} of them, then at any one moment some of the protons will be interacting to produce a deuterium nucleus. This the story of a statistical tail wagging the dog. Once a deuterium nucleus has formed the process moves on much more rapidly, and within a matter of seconds the ^2H will interact with a proton to produce helium-3: $^2H+p \rightarrow {}^3He+\gamma$. And finally, the process ends with the interaction of two helium-3 nuclei to produce helium-4, with its nucleus appropriately composed of two protons and two neutrons: $^3He+{}^3He \rightarrow {}^4He+2p$. Some energy is liberated in each of the steps just outlined, in what is called the proton-proton chain, although the majority of energy is actually liberated in the first, deuterium-forming step.

At the same time that Bethe and Critchfield were describing the machinations of the proton-proton chain, Carl von Weizsäcker (and independently Bethe) developed an alternative method by which hydrogen could be converted into helium. In this second process a catalytic cycle was envisioned. In this case a proton interacts with a carbon-12 nucleus to produce nitrogen-13, which after a series of β-decays and additional proton captures results in the production of nitrogen-15. The final step in this process, called the CN cycle, requires the nitrogen-15 nucleus to capture a proton and then decay into carbon-12 via the emission of a helium-4 nucleus.

Since the sequence begins and ends with a carbon-12 nucleus, the net result of the CN cycle is the transformation of 4 protons in to one helium-4 nucleus with the liberation of energy. Additional branches have been identified within both the proton–proton chain (so-called PPII and PPIII) and the CN cycle (the CNO cycle), but the key point is that as of the late 1930s at least the mechanisms by which stars generate energy within their interiors had been identified. It transpires that low mass stars up to about 1.5 $M_\odot$ are powered by proton–proton chain reactions, while more massive stars are powered by the CNO cycle.

At the time that they were first developed it could be argued that the more natural mechanism was that of the proton–proton chain, since it involved just protons, and it was then well established that stars were mostly hydrogen. The CN and CNO cycles require the pre-existence of carbon-12, and in the late 1930s it was far from clear where this element might have come from. Indeed, to be fair, it was not then clear where any of the atoms in the Periodic Table actually came from. Were all the atomic elements always there in space, or were they generated by physical processes not, as of then, envisioned? The answer to this problem was eventually provided for in the late 1950s.

To conclude this rather lengthy technical section we have now identified the conditions for the attainment of stardom. Once nuclear fusion reactions are initiated in the interior of a star it need no longer draw upon the release of gravitational energy to stay hot within its interior, and it settles down to a long-lived, stable, dynamical equilibrium, phase. Indeed, the newly formed star has settled onto the main sequence.

B²FH

The progress of science is not entirely built on the struggle to bring new ideas to fruition, or by observing a totally new phenomena. Oftentimes progress is made through synthesis. At such times a whole series of ideas, built up independently by many observers across a number of different research fields, are brought together to reveal a unity of concept.

Such a synthesis of ideas and concepts was published in a paper by Elizabeth (Margaret) Burbidge, Geoffrey Burbidge, William Fowler and Fred Hoyle in 1957. Generally known as B²FH after the surname initials of the authors, this paper grew out of an earlier *Science* magazine publication, *Origin of the Elements in Stars*, published by HFB² on October 5, 1956. Both papers set out to demonstrate that "the [atomic] elements have been and are still being synthesized in stars." Both papers outline what is now the fully accepted concept that all of the atomic elements, essentially from boron [14] all the way through to uranium, are built up step by step within the nuclear burning regions of stars and in supernovae.

Although certainly relying on many speculative ideas, both of these papers, especially B²FH, had a profound effect on subsequent astronomical research and cosmological theory. Indeed, these papers essentially established the field of stellar nucleosynthesis.

Published in the journal *Reviews of Modern Physics* B²FH begins with a quotation from Shakespeare's *King Lear*: "It is the stars, the stars above us, govern our conditions," and while Shakespeare intended an astrological meaning, it is, in modern terms, an apt quotation with respect to the origin of life in the cosmos. The B²FH paper is long, 108 pages long, but bristles with profound insight. In terms of importance the SAO/NASA Astrophysical Data System indicates that as of June 2016 the B²FH has been referenced 1844 times in other research papers.[8]

Although the citation level of a paper is not an especially good indicator of how influential the work has been, it certainly serves as a guide to its impact upon the thinking of other researchers. At the heart of B²FH is the idea that "all prompt nuclear processes can be described as the shuffling of protons and neutrons into a variety of nucleonic packs… [and that]…nuclear processes plus the slow beta reactions make it possible in principle to transmute any one type of nuclear material into any other." By such statements the authors of B²FH envisioned the generation of all atomic nuclei beyond helium through stellar fusion reactions, the capture of neutrons and α-particles (helium nuclei) by atomic nuclei, and transformations associated with nuclear decay.

The process runs step by step and is predicated on the notion that stars are self-governing mechanisms. The temperature within the central core of a star is adjusted so that the energy outflow at the surface of the star is exactly counterbalanced by the generation of energy by fusion reactions in the core. Hydrogen fusion reactions run at a temperature of about 10 million degrees, followed by helium fusion via the triple alpha reaction, which runs at a temperature of about 100 million degrees and so on upwards through carbon burning all the way to iron. Between the exhaustion of one fusion fuel and the onset of the next, a star's central core must heat up, and this increase in temperature is

---

[8] By way of comparisons the Astrophysical Data System indicates that Stephen Hawking's *A Brief History of Time* (published in 1988) has been reference 108 times. Hawking's first paper on black hole radiation, however, has been referenced 5369 times. The author's best citation count is 77, which speaks volumes!

brought about by shrinking. In this adjustment phase, recall potential gravitational energy is converted into thermal energy. For stars up to about 8 times the mass of the Sun, the central cores are incapable of configuring density and temperatures that allow fusion reactions beyond the triple-alpha process. For the stars more than 8 times the mass of the Sun, however, the fusion process halts at iron (which has the most strongly bound atomic nucleus), since reactions beyond iron are endothermic, meaning that the star cannot generate energy from any further conversions.

At this stage it is the incredible and very rapid release of energy during a supernova explosion that is needed to synthesis the remaining elements up to uranium. These latter reactions were first described in detail by Alastair Cameron (then at Atomic Energy of Canada Limited, later at Harvard University) in a research paper appearing in the *Publications of the Astronomical Society of the Pacific* in the same year (1957) as B^{2}FH. In this paper Cameron essentially outlined the domain of what has become known as explosive nucleosynthesis, where element transmutation takes place in a neutron-rich environment.

Indeed, B^{2}FH and Cameron realized that not every atomic element is directly synthesized through fusion reactions, and it was realized that additional atomic processes are required to complete the full complement of the Periodic Table. In fact, there are quite a few additional nuclei-building mechanisms available, but the main ones invoked are the so-called s- and r-processes involving neutron capture. Here the s- and r-stand for slow and rapid, and the relative measure for the speed is that of beta-decay. In the latter process a neutron spontaneously transforms into a proton and ejects an electron from the nucleus. There is also an inverse beta decay in which a proton transforms into a neutron and ejects a positron from the nucleus.

The point, however, is that some nuclei are built up through neutron capture on a timescale compared to the beta decay process. These are the s-process elements. Other elements are built up through neutron capture on a timescale short compared to the beta decay. These are the r-process elements. There is also a p-process involving proton capture, and an α-process in which nuclei are built up through the capture of α-particles. Current theory indicates that low mass stars are responsible for generating most of the C, N, and s-process isotopes found in the interstellar medium, while massive stars and supernovae are largely responsible for generating the O and r-process isotopes and the elements beyond iron all the way to uranium.

In 1997 the "Reviews of Modern Physics" was published, under the lead authorship of George Wallerstein (University of Washington), a retrospective article on the state of knowledge 40 years on from the appearance of B^{2}FH. This latter review has the more awkward surname acronym of WIPB^{3}H^{2}CKS^{2}TML (there are 15 authors). It summarized the great progress that had been made in gathering key observational data, computational modeling, and the understanding of atomic transition probabilities, but find, nonetheless, that while there are still gaps in our knowledge, the basic premise behind the B^{2}FH and Cameron papers remains true. The stars are indeed the great cauldrons in which all the elements beyond those derived by primordial nucleosynthesis are produced. It is the stars that are slowly transforming the chemical makeup of the universe, and it is through the inner workings of the stars that life has been able to come about.

American astronomer Carl Sagan once wisely quipped that to the make an apple pie from scratch, one has first to create a universe. To this might now be added that the universe must also be one in which stars like our Sun can form [15]. Importantly, the B^{2}FH paper not only outlined the way in which the chemical elements can be generated, its authors also realized that to explain the observed chemical abundances relative to hydrogen, there must be a interchange process at work, with the material of the interstellar medium being recycled and further processed by each successive generation of stars, each dying generation returning enhanced material (that is, elements other than hydrogen and helium) back into the ISM during its planetary nebulae and supernovae end phases. Life in all of its wonderful forms, from bacteria to barnacles, baboons to blue whales, is only possible today because myriad stars have formed and died in the distant past. What an incredible alchemy of change, transformation, and evolution.

Both Cameron and B²FH highlighted supernovae as important engines of chemical transformation. Not only is the high energy and neutron-rich environment of the supernova envelope required to synthesize nuclei, but so, too, is the explosive ejection of material into the surrounding interstellar medium vital for the mixing process. Indeed, the end phases of all stars, either as planetary nebulae or supernovae, are involved in returning and mixing nuclear processed material into the interstellar medium.

As time marches indomitably by, the very chemical makeup of the galaxy, as expressed through the composition of the interstellar medium, changes. This change in galactic composition is evidenced by the various populations of stars identified by Walter Baade. Population I stars are metal rich, by which it is meant that the mass fraction of elements other than hydrogen and helium is high. In contrast Population II stars are metal poor. Linking the metallicity of a star to the environment out of which it formed reveals a time, or age, progression. Population II stars must be older than Population I stars since they have relatively fewer nuclear processed elements within their interiors.

An even older set of stars, Population III stars, has additionally been invoked to explain the fact that Population II stars are not completely devoid of heavy elements. Population III stars are the first-born stars, and they are taken as being composed of just hydrogen and helium. This pure hydrogen/helium composition has important consequences with respect to the mass range of stars that can form (see below). These time changes in stellar population characteristics and the chemical abundance of the interstellar medium can be modeled on the grand galactic scale by convolving models of the star formation rate, the initial stellar mass function (which dictates how many stars are formed in a given mass range), and a set of stellar evolution models, the end fate of which is determined by the initial mass.

One of the early pioneers in this latter field of study was Beatrice Tinsley (Yale University[9]) who, during the 1970s, developed detailed models to describe the time evolution of galactic chemical composition. Essentially Tinsley's idea was to develop a time-step model, with the composition at a specific galactic location at time $t + \Delta t$ being determined by the composition at that location at time t adjusted for the amount of new star formation and the progress of stellar evolution (specifically the end phase evolution) in the time interval Δt. In this manner, a picture of the time evolution of galactic chemical composition can be built up step by step. Interestingly, what the observations and the galactic models indicate is that the star formation rate must have been much higher in the distant past. Such models also indicate where and when in the galaxy the conditions become suitable for the stable development of life (as we shall see in Chap. 6).

---

[9] Tinsely was, in fact, the very first female professor of astronomy at Yale, taking up the post in 1978.

The Engines of Evolution

With the advent of fast electronic computation in the 1960s it became possible to explore large swaths of parameter space relating to stellar evolution. In this manner the journey of a star, from the main sequence to its end phase, could be studied and traced within the HR diagram. The mass domain of stardom could also be explored, and the calculations indicate that the smallest mass M_{min} for which the central temperature and density allow for the onset of hydrogen fusion reactions is 0.08 $M_\odot$.

Objects below this mass can, and indeed are observed to, form, but they have no long-running internal energy source to replace the energy radiated into space at their surface. Such brown dwarfs, once formed, are on a continual cooling off sequence. The upper mass limit to stardom is less clear. The problem for massive stars is that their high internal temperature means that radiation pressure becomes very important, and this limits not only the accretion process but drives strong atmospheric winds. Additionally, when radiation pressure dominates, the star hovers on the brink of instability, adopting a maximum luminosity, the so-called Eddington limit (after his pioneering analysis of such extreme stars [16]).

The present upper mass limit M_{max} for stars is thought to be about 125 $M_\odot$. Numerical experimentation with stellar evolution codes has further revealed a third critical stellar mass, with M_{crit} being 8 $M_\odot$. For stars with initial masses within the range $M_{min} < M < M_{crit}$ the end phase of stellar evolution is the production of a planetary nebula and a white dwarf remnant. For stars in the mass range $M_{crit} < M < M_{max}$, the end phase is supernova disruption, with the production of a neutron star if $M < 50\,M_\odot$ and a black hole if $50\,M_\odot < M < M_{max}$. The three channels of stellar evolution are illustrated in flowchart fashion in Fig. 5.10. Importantly for our narrative is that the end phases of all stars from M_{min} to M_{max} results in the return of star-processed material to the interstellar medium.

The evolution of the Sun from its present age (being 4.5 billion years old) to its late planetary nebula phase is illustrated in Fig. 5.11. This diagram illustrates the changes in the Sun's radius, luminosity, and surface temperature. At the present time the Sun is a main sequence star generating internal energy through the conversion of hydrogen into helium within its core. As described earlier the Sun has a main sequence (nuclear burning) lifetime of about 10 billion years; it is therefore about middle-aged. As the Sun continues to age, however, its luminosity and radius will begin to increase, while its surface temperature will decrease. At this stage it is evolving along the so-called red giant branch (RGB). At the tip of the RGB, in about 7.5 billion years from now, the Sun will be 160 times larger, 2300 times more luminous, but 2500 degrees cooler than at present.

It is also at this stage that helium burning, through the triple-alpha reaction, begins in its core. This process begins rapidly resulting in what is called the helium flash (HeF in Fig. 5.11). Once initiated, however, the Sun will settle down to steady helium burning (3He in the figure) at a much lower luminosity and size to that exhibited at the red giant tip. At this stage the Sun will have found its second long-lived fuel supply, but as with all finite things, the helium is gradually converted into carbon. The helium-burning lifetime of the Sun is much shorter than that of its main sequence lifetime, being some 100 million years, since the fuel supply of helium is smaller than that of the hydrogen in the main sequence phase and the luminosity during helium burning is much higher.

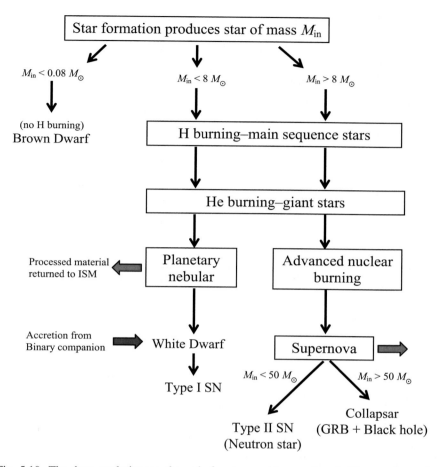

Fig. 5.10 The three evolutionary channels for stars and brown dwarfs. The star formation process determines the initial mass via the initial mass function (IMF). *Red arrows* indicate stages at which processed material is returned to the interstellar medium. The *blue arrow* indicates mass accretion from a binary companion

As the helium is consumed in the Sun's central core it begins to climb what is called the asymptotic giant branch (AGB). Once again its radius and luminosity begin to increase dramatically. It is towards the top of the AGB that the physics becomes fiendishly complicated, and the Sun will develop multiple nuclear burning shells in its interior. Above a now helium-exhausted core there will be a helium-burning shell, and above this there will be a thin shell in which hydrogen burning is taking place. These shells will turn on and off according to a whole slew of complex interactions, and the Sun will undergo what are called thermal pulsations, literally expanding and contracting on timescales of order days to many months. At this stage the Sun will become a Mira variable.

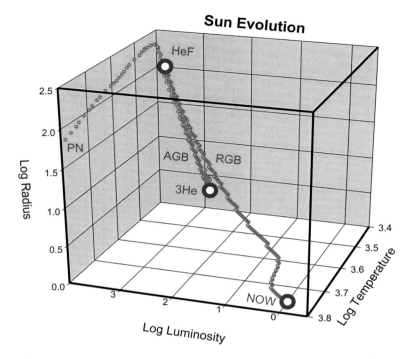

Fig. 5.11   A 3-D HR diagram of the Sun's evolution. See text for details

During these pulses various dredge-up phases will be initiated within the Sun, with nuclear processed material being brought towards its surface by deep convective transport. The onset and extent of the dredge-up phases are highly sensitive to model assumptions, and they are still not fully understood. The key point, however, is that processed material that was otherwise buried deep within the Sun is brought towards its outer layers, from where it can be returned to the interstellar medium during mass losing episodes and during the formation of a planetary nebula (PN).

The wind-driven mass loss from Mira variables are observed to be very high, amounting to some 10^{-5}–10^{-4} solar masses per year, although the details of how such "super-winds" come about is far from clear. What does appear to be clear, however, is that the pulsations drive the formation of shock waves within the outer layers of the star, and that within the same region of the envelope carbon dust grains form by condensation. Indeed, these dust grains, once formed, are picked up by radiation pressure and driven into the interstellar medium. It is these same dust grains that, once diffused through the interstellar medium, produce interstellar extinction, the zone of avoidance, reflection nebulae, and allow for the formation of molecular clouds.

The instability leading to the planetary nebula phase of the Sun is again not well understood, but it appears that in the time that the Sun evolves from its early helium burning phase (3He) to its planetary nebula onset stage it will have lost nearly 40% of its mass.

It is the processed material that goes into the planetary nebula that is ultimately returned to the interstellar medium. The core of the Sun, the remaining 60% following the PN phase, will evolve towards a white dwarf configuration, and thereafter it will slowly cool off until it becomes a zero temperature, zero luminosity, but stable against gravitational collapse, black dwarf.

Although low mass stars have a relatively low-key, quiescent death, massive stars signal their demise in a blaze of glory. As indicated by Fig. 5.10, massive stars can configure their interiors to enable fusion reactions beyond helium burning to take place. First a series of carbon fusion reactions occur, producing such elements as neon, sodium, and magnesium. In short supply, however, such carbon burning can only proceed for a few hundreds of years—small fry in terms of the much lengthier main sequence phase.

With the exhaustion of carbon, neon photodisintegration can take place, producing oxygen, and then the oxygen can undergo fusion reactions for perhaps just a few months to produce silicon. Ultimately silicon can fuse with helium to produce sulfur and eventually iron—a process that can run for about 1 day.

Now the end is near. Every available fuel for fusion reactions has now been consumed within the star, and what is left is a core predominantly composed of iron nuclei. Here the fusion process stops, since iron is the most tightly bound atomic nucleus. The star has now exhausted all of its options with respect to generating energy through fusion reactions. Surrounding the iron core is a series of shells rich in the ashes from earlier fusion reaction stages. Around the iron core is a shell rich in silicon nuclei; this in turn is surrounded by a shell rich in oxygen nuclei, then a shell of neon, carbon, helium, and finally an envelope rich in hydrogen nuclei. The core temperature is now several billion Kelvin, and the density is around 10^{12} kg/m^3. For all this, however, the core is supported by the pressure resulting from degenerate electrons, and it will remain stable against gravitational collapse provided its mass does not build up beyond about 1.4 solar masses (the so-called Chandrasekhar limit).

Above the Chandrasekhar mass limit the electrons can no longer hold gravity at bay, and the core rapidly collapses to produce a neutron star. This results in the formation of what is called a Type II supernova. The collapse time t_{coll} is very short, being about 0.05 s. In this blink-of-the-eye time gap the core shrinks from about the size of Earth to the size of a neutron star, which is about 20 km across. The liberation of gravitational energy in this collapse is enormous, amounting to some $\Delta E \approx 5 \times 10^{46}$ J. This is the same amount of energy that would be released by the complete annihilation of some 93,000 Earths. The luminosity of the core collapse will be of order the collapse energy divided by the collapse time: $L_{coll} \approx \Delta E / t_{coll} \approx 10^{48}$ J/s $\approx 2.5 \times 10^{21}$ L$_\odot$.

It is difficult, even by astronomy standards, to express just how much energy is released during the brief time interval of core collapse. The luminosity of the entire Milky Way Galaxy is estimated to be of order 2×10^{10} L$_\odot$, so, accordingly, in the fraction of a second it takes the iron core to collapse to a neutron star the energy released is equivalent to the energy radiated into space by 100 billion galaxies in 1 s. If we recall the Eddington number from earlier in this book, we have seen that our observable universe contains some 65 billion galaxies. Accordingly, a single core collapse supernova event, in just 0.05 s, releases as much energy as the entire observable universe of galaxies radiates in the form of starlight in 2 s.

The vast majority of the energy created during core collapse is actually carried away by neutrinos, which are illusive, weakly interacting, and almost impossible to detect particles. The Sun continuously generates neutrinos in its core, for example, and of order 10^{15} (a thousand million million) neutrinos pass through our bodies (having an area of about 1 sq. m) every second of every day (that's 86 million million million neutrinos per day).

The Sudbury Neutrino Observatory (SNO), located 2 km underground, uses 1000 m. tons of deuterated water (D_2O), contained within a 12-m diameter acrylic sphere, to detect neutrinos. This arrangement makes for about 10^{31} target atoms with which solar neutrinos might interact, and during a typical 24 hour period, of the 10^{22} solar neutrinos that pass through the detector just 10 interaction events are recoded. Neutrinos, to put it mildly, are shy. For all of their retiring nature, however, neutrinos were positively detected in the aftermath of the supernova event SN 1987A, which occurred in the Large Magellanic Cloud 168,000 light years away on February 24, 1987.

The vast number of neutrinos generate during core collapse are produced by a whole slew of interactions. One channel for their production is neutronization, where the protons and electrons within the collapsing core combine through the reactions $p + e^- \rightarrow n + \nu$. Other pathways for generating neutrinos include electron and positron annihilation $e^+ + e^- \rightarrow \nu + \nu'$ (where ν' denotes an anti-neutrino), plasmon decay: $\gamma \rightarrow \nu + \nu'$, and photo-annihilation: $e^- + \gamma \rightarrow e^- + \nu + \nu'$.

It is the vast number of neutrinos and the extremely high density of the collapsing region that ultimately causes an explosion that reverses the collapse of the outer layers, thereafter driving material into interstellar space at speeds in excess of 1000 km per second. Only about 1 % of the energy released during core collapse goes into the kinetic energy of the expanding envelope, and less still, about 0.01 % of the energy is radiated in the form of visible light. It is within the expanding outer layers that the r-process has its brief chance to build up atomic nuclei beyond that of iron.

In the case of SN 1987A the progenitor star (Sanduleak −69° 202) had an estimated initial mass of about 20 $M_\odot$, and it is thought that perhaps 18 solar masses worth of material was returned into the surrounding interstellar medium. Of this returned material, perhaps 2.5 $M_\odot$ was in the form of elements other than hydrogen and helium. A 1.5 $M_\odot$ neutron star probably formed during the core collapse leading to SN 1987A, but no direct evidence for this, in the form of a pulsar, for example, has so far been found. Some 500 days after the initial detection of SN 1987A the first signs of dust grains being produced in the expanding envelope were reported, and some 24 years on, using Herschel Space Observatory data, it has been estimated that perhaps some 0.5 $M_\odot$ of dust had been produced by 2011 (Fig. 5.12).

A crude estimate of how much interstellar material has been recycled through supernovae since the galaxy formed can be made upon the evaluation of how many occur per year at the present time. Only 8 supernovae have been observed within the Milky Way Galaxy during the past 2000 years, which is on average 1 every 250 years, but this number is much smaller than the true rate because of the absorption of starlight by interstellar dust. Observations of external galaxies indicate, however, that spiral galaxies like our own have supernova rates of about 1 in every 50 years.

Fig. 5.12 Supernova 1987A. The supernova is visible as the very bright star to the middle right in the image. The star formation region corresponding to the Tarantula Nebula is visible in the *upper left* of the image. Image courtesy of ESO

If the supernova rate is assumed to be constant (it actually isn't, as we'll see later), the age of the galaxy is taken as being 13 billion years, and given that a typical supernova progenitor star has a mass of 12 $M_\odot$ with 10 $M_\odot$ being returned to the interstellar medium at the time of core collapse, then of order $(13 \times 10^9/50) \times 10 = 2.6$ billion solar masses of interstellar matter has passed through at least one supernova event. In turn this supernova processing will have generated some 300 million solar masses worth of nuclear processed material.

With the time evolution of stars and their end phases (reasonably well) understood it is possible to trace the time history of the interstellar medium and the evolution of galactic chemical composition. The cycle is exactly that outlined earlier in this chapter. At each time-step forward there are new stars being born, removing material from the interstellar medium, and old stars going through their end phases returning nuclear processed material to the interstellar medium. As the time-steps add up, more and more interstellar matter will have been processed through stars of one mass or another, and the composition of the interstellar medium will become increasingly metal-rich in atoms other than hydrogen and helium.

The first step in the process, however, is that of primordial nucleosynthesis, and this describes the creation of hydrogen, helium, lithium, beryllium, and boron in the first tens of minutes after the universe came into existence at the time of the Big Bang some 13.7 billion years ago. From this primordial hydrogen and helium mixture the first generation (Population III) stars formed, and within a few millions of years the first supernovae began to explode. The first iteration of recycling will have then taken place, and the rest is literally history.

No recycling process is ever 100 % efficient, and while the composition of the ISM is slowly being changed over time it is also being depleted of star-building material. The reasons for this material loss are indicated in Fig. 5.13, and correspond to the mass sinks resulting from the formation of white dwarfs, neutron stars, and black holes. These objects, if left alone, will remain stable for all time, and the matter they contain will no longer be available for incorporation into new stars. Ultimately star formation will cease, and the galaxies will slowly begin to blink out from all visibility. For all this, however, as time goes by, the interstellar medium will become more and more enhanced in those elements beyond helium, and this, of course, has consequences.

The long term fate of the universe has been described in great theoretical detail by Fred Adams and Gregory Laughlin in an excellent paper published in the April 1, 1997, issue of *Reviews of Modern Physics*. Building upon many classical results, Adams and Laughlin outline in their review how stars will behave as they become increasingly metal-rich. Indeed, it turns out that, as a result of increasing metallicity, low mass stars become less luminous at a given mass, while the most massive stars become more luminous at a given mass. This result comes about mostly because of the complex way in which the opacity (which measures how much the stellar gas hinders the passage of electromagnetic radiation through it) is affected by changes in the metallicity. Furthermore, the actual size of main sequence stars increases, over all mass ranges, as the metallicity increases, and this results in the stars having higher surface temperatures.

The main consequence of these changes, with respect to the main sequence lifetime, is such that stars will spend less time in the hydrogen-burning phase and therefore will be less long lived. Additionally, numerical experimentation on stellar models indicates that the larger the metallicity the lower are the upper and lower mass limits for stardom.

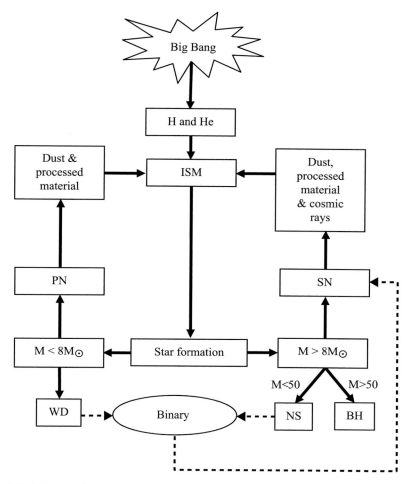

Fig. 5.13 The recycling loop from the ISM to stars and back to the ISM. The first step in the loop is the formation of hydrogen and helium through primordial nucleosynthesis. This is the raw star-making material of the universe. There will also be an early ISM enhancement phase relating to the formation of predominantly massive Population III stars. The star formation box splits two ways according to the initial mass being greater than or less than 8 times the mass of the Sun. The *dotted line* from the 'binary' ellipse indicates the potential outcome of mass accretion in a binary system, resulting in otherwise stable white dwarf or neutron stars being pushed over their critical mass limits or coalescing to produce Type Ia supernovae in the case of white dwarfs, and kilonovae (A kilonova (also called r-process macronova) occurs when either two neutron stars coalesce or a neutron star and a black hole merge. They are associated with short gamma-ray bursts, and they are thought to be responsible for the production of most of the r-process elements in the universe. They are also touted as being strong candidate sources for detection with gravitational wave instruments such as LIGO [17]) in the case of two coalescing neutron stars. *ISM* interstellar medium, *PN* planetary nebula phase, *SN* supernova phase, *WD* white dwarf, *NS* neutron star, *BH* black hole

The upper mass limit, which is set, you'll recall, by the Eddington luminosity [16], is achieved at a lower mass value, and in the extreme limit Adams and Laughlin find that the maximum mass of a star will be reduced to of order 30 $M_\odot$, some 5 times smaller than the upper mass limit at the present time.

The lower upper-mass limit to stardom has additional affects with respect to the end mass of a star prior to its supernova detonation. Specifically, the mass loss via stellar winds increases as the metallicity increases, and consequently the stars undergoing supernova detonation will have smaller masses. This in turn begins to change the likely detonation mechanism away from that of an iron-rich core collapse to that of electron capture in a degenerate core rich in oxygen, neon, and magnesium. Rather than the core collapsing once its mass exceeds the Chandrasekhar limit, set by the degeneracy of electrons, the collapse is initiated by the neon and magnesium actually capturing electrons, thereby removing the star's pressure support mechanism. The change in the core collapse mechanism results in smaller amounts of nuclear processed material being returned to the interstellar medium, and it will also mean the end of stellar-mass black hole formation, since the production of a neutron star is now much more likely.

The lower mass limit for stardom is also reduced by an increase in the metallicity. And again, Adams and Laughlin suggest that when the metallicity reaches a level twice that of the current solar value, objects with masses as small as 0.04 $M_\odot$, half that of the present lower mass limit, will be able to begin burning hydrogen within the cores. These metal-rich lowest mass stars, Adams and Laughlin further note, will be decidedly strange objects, and they call them "frozen stars." Indeed, although such low mass objects will be converting hydrogen into helium in the cores, though at a minimally slow rate, their surfaces will be cool, having temperatures of order 275 K—that is, the freezing point temperature of water. These frozen stars will have luminosities perhaps a thousand times smaller than the current lowest mass stars and will be able to sustain fusion reactions within the core for many trillions of years.

As the ISM recycling loop-counter continues to tick forwards in time, the stars will change, and indeed, star formation will change with the IMF favoring more and more the formation of lower mass stars. There will also be consequences for planet formation as the metallicity increases.

Present survey data indicates that the likelihood of exoplanets being found in orbit around a star increases with the increasing metallicity of the host star. Accordingly, as the interstellar medium becomes increasingly metal-rich, so planet formation should become more and more prevalent in low mass stars. In a 2015 study concerning the future of planet formation within the galaxy and larger universe, Peter Behroozi and Molly Peeples (both at the Space Telescope Science Institute, Baltimore) argue that planet building will continue for a least the next several trillion years. Additionally, however, Behroozi and Peeples suggest that 80 % of the presently existing Earth-like planets in our galaxy formed before the birth of the Solar System. However, they additionally argue that, taking future star formation into account, of all the Earth-like planets that will ever exist in our Milky Way Galaxy, Earth formed prior to 61 % of them coming into existence. The Milky Way has yet to generate the majority of its planets. Additionally, Behroozi and Peeples find that of all of the Earth-like planets that will ever exist in the observable universe, 92 % of them, have yet to form.

First and Last

The star formation rate (SFR) was discussed earlier, and it was indicated that the SFR in a given galaxy could be derived from its integrated light. For distant galaxies this measure is built around the SFR of the brightest (and hottest) O spectral-type stars, the fainter, more numerous low mass stars not being visible in distant galaxies. Using, therefore, a measure of the SFR for the brightest, most massive stars, assuming that the initial mass function (IMF) is constant over time and in all star-forming environments, and adopting a series of stellar population models, it is possible to deduce the star formation history of the universe.

This unraveling of history is possible by combining the SFR data with a measure of the distance to the galaxies being studied, since the further away a galaxy is, the longer it has taken its starlight to reach us and accordingly the greater is the look back in time. Figure 5.14 shows the unraveled history of star formation in the universe. Indeed, the analyzed light from the most distant galaxies indicates that star formation was much more vigorous in the young universe, and that star formation was at its peak about 10 billion years ago, when the universe was some 3–3.5 billion years old.

After the peak the universal SFR has been declining exponentially, falling by a factor of about 3 every 3.9 billion years. The star formation rate at the present time is about 10 times smaller than it was at the time of peak star production. Indeed, it is estimated that about half of the stars observed today were formed before the universe was 4 billion years old. The reason for this higher star formation rate in the distant past is thought to relate to the fact that there were more galaxy-galaxy interactions in the past, and the fact that galaxies took several billions of years to acquire their full gas-mass complement. As we move into the distant future, however, fewer and fewer stars will be produced, and current models suggest that the majority of all the stars that will ever form have already done so. This being said, star formation, though at an ever dwindling rate, will continue for many hundreds of billions of years yet to come.

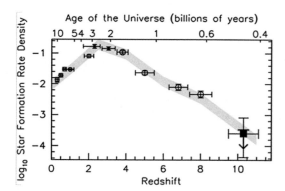

Fig. 5.14   The history of star formation as measured at various look-back times. This diagram is based upon the integrated starlight of 221,373 individual galaxies at wavelengths between 0.1 (far-UV) to 500 (far-IR) microns. Image courtesy of Bouwens/firstgalaxies.org

The very first stars in the universe appeared at a time when it was less than half a billion years old. These first-born stars constitute the so-called Population III objects that have been added to Baade's original classification scheme of stellar populations. The important distinguishing property of Population III stars, at least the very first ones, is that they contain no chemical elements other than hydrogen and helium; they are zero metals stars. Indeed, they are made of the pure hydrogen and helium that was forged during the brief few minutes of primordial nucleosynthesis following the Big Bang moment of creation.

Not only are these the first born stars, but they are also the first lamps to illuminate the universe. The population III stars banished the 350 million year interval, called the dark ages, that followed the decoupling of radiation and matter 380,000 years after the Big Bang. Indeed, this latter moment of decoupling was the time in which the texture presently observed in the cosmic microwave background was imprinted.

Not only did the Population III stars banish the darkness that pervaded the early universe, they also re-ionized much of its constituent matter. This re-ionization epoch fundamentally changed the structure of the universe, and it represents the very first time that ordinary (baryonic) matter had any direct influence on events occurring in the universe,[10] taking it from a neutral one, where every atomic nucleus had its appropriate electron complement, to the plasma dominated one we see today. Indeed, the super-low density intergalactic medium, containing about one atom per cubic meter, shines brightly at X-ray wavelengths with a thermal temperature of many millions of degrees.

The formation of the first stars will have run in essentially the same way as described above, but, for all this, there must also have been fundamental differences in the basic evolutionary process. First, when the first stars formed, there were no galaxies, no interstellar dust grains, and no molecular clouds. The first-born stars were composed entirely of hydrogen and helium, and they were most likely predominantly massive stars.

Although the details are far from clear, the latter statement on stellar bulk is based on the fact that it is only through the shielding effects afforded by interstellar dust that molecular clouds can form and then remain cool; without dust the first star-forming gas clouds must have been warm. There was possibly some H_2 in the primordial star-forming gas clouds, formed as a result of gas phase interactions, but even so, the first Population III stars probably formed in environments at least ten times hotter than that observed in present-day (Population I) star forming molecular cloud cores.

Importantly with respect to making stars, however, the Jeans mass in a warm gas cloud is larger than that in a cooler one, and consequently it is thought that the first zero-metal Population III stars were true behemoths, with masses possibly running as high as several hundred to possibly even thousands of solar masses. The lifetime of such massive stars is measured in just a few millions of years, and accordingly they would soon undergo supernova disruption, and thereby begin to seed their surroundings with nuclear processed material and dust. This would complete the first loop in the interstellar recycling network. These massive first stars would also endow the universe with a plethora of stellar mass black holes [18] and presumably gravitational wave generating coalescent events, the likes of which would eventually be detected by LIGO [17] or its successors.

---

[10] Prior to re-ionization the entire structure and evolution of the universe was determined by its dark matter content alone.

The large characteristic mass associated with the first stars dictates that they should be very luminous, but the unambiguous detection of Population III starlight has proved illusive. Such stars are not visible in the Hubble Deep Field, nor do they appear in Hubble's ultra-deep field or extreme deep field images. The advanced technology and greater light grasp of the James Webb Space Telescope, however, should potentially find clear evidence of the first-born stars, but nature has always found a way of upsetting even the best laid out of theories, so at the moment only time will tell what will be seen of this early star-forming epoch.

Given that the first stars began to form in the universe when it was just half a billion years old, one might additionally ask, when did the first planets form? Surprisingly, perhaps, the answer to this question may be as early as a few tens of millions of years after the very first stars. Some planetary real estate may be decidedly ancient.

To make a planet, either terrestrial or Jovian, one needs elements other than hydrogen and helium; indeed, metals are essential, and these latter elements will only be available after the very first Population III stars have evolved and undergone supernova disruption. Second generation Population III and the oldest Population II stars will contain the requisite metals to start forming planets, although the first planets to emerge may well be very different in nature to those found in our Solar System.

The most metal poor stars in the Milky Way Galaxy, Baade's Population II stars, are predominantly found in the halo of the galaxy, within globular clusters, and in the galactic core. Those Population II stars within the halo have large elliptical orbits around the galactic center that carry them well above the galactic plane, and they are often detected as a result of their large spatial velocity. Of the oldest and most metal poor stars known, that of SMSS J031300.36-670839.3 holds the present record. Located some 6000 light years from the Sun, and estimated to be some 13.6 billion years old, this star was described by Stefan Keller (Mount Stromlo Observatory, Australia) and coworkers in the online February 9, 2014, issue of *Nature*.

SMSS J031300.36-670839.3 is remarkable, and though it is not a first born Population III star, it may well be a second generation star. The spectrum reveals no iron absorption lines to the level of detection (that is, the abundance is smaller than one-ten millionth that of the Sun), and it is suggested by Keller et al. that elements other than hydrogen that are visible in its spectrum are the dusting of metal debris from just one Population III supernova event. Many extremely metal poor stars are known, but one subgroup of these, the so-called CH stars, has proved particularly interesting with respect to recent astrobiological speculation.

First described by Philip Keenan (Ohio State University) in 1942 the CH stars are distinguished by their spectra; although indicating a very low iron abundance, they show strong molecular bands due to the CH molecule (methylidyne). CH stars are found in globular clusters and in the galactic halo, and have more recently been subsumed into the carbon enhanced metal poor (CEMP) group of stars. The CEMP stars were first studied in detail by Timothy Beers (Michigan State University), George Preston and Stephen Shectman in 1992, and several sub-groups are now recognized.

The observational data indicates that the CEMP stars are the more extreme versions of the CH stars, and that they become more frequent in number with distance away from the galactic plane and with decreasing metallicity. There are probably several mechanisms responsible for the formation of CEMP stars (mass transfer in a binary system or AGB dredge up), but Beers and coworkers argue that they are additionally stars on the classification boundary between late-forming Population III and early-forming Population II stars, and that their spectra show the record of the nucelosynthesis products of the first wave of Population III supernovae.

One particular group of CEMP stars, the so-called CEMP-no stars (the 'no' indicating that such stars have a very low abundance but normal pattern of heavy elements) have recently been highlighted by Natalie Mashian and Abraham Loeb (both of the University of Michigan) as possible candidates for hosting planets made of carbides and graphite (This is in contrast to the silicates and metals typically associated with terrestrial planets). Indeed, Mashian completed her doctoral thesis on this topic in 2016, and has specifically argued [19] that the carbon-rich properties of CEMP-no stars may extend to that of their formative disks, the disks being enhanced in the carbon dust grains produced in the first few waves of Population III supernovae ejecta.

By constructing models of carbon planets, Mashian and Loeb found that a maximum size, of about 4.5 times that of Earth, is reached at a mass of about 1000 M_{Earth}, while a 1 M_{Earth} carbon planet is about the same size as Earth. Planets of this size are well within the detection limits of modern transiting exoplanet surveys. Importantly, Mashian and Loeb point out that a carbon planet could be distinguished from a water or terrestrial planet by the presence of atmospheric CO, CH_4, and hydrocarbons, and an absence of such molecules as H_2O, CO_2, O_2 and O.

The idea that there might be such objects as carbon planets was first put forth by Marc Kuchner (Princeton University) and Sara Seager (Carnegie Institution of Washington) in 2005, where it was suggested that they might be found in orbit around highly evolved objects such as pulsars and low-mass white dwarfs. A first-born, pure carbon planet, however, will have characteristics strikingly different than those displayed by a similar mass terrestrial planet; for example, pure carbon planets could not support an atmosphere-protecting magnetic field (there is no hot-iron core), and even if an atmosphere were present there would be no surface water (since the oxygen abundance is essentially zero). None of these latter issues are especially conducive to the idea that life (at least as we know it) might have first evolved on carbon planets, but nature has always found unexpected ways of surprising us.

Indeed, it is interesting to ponder this question. Given that planets may have existed in the universe since the first few hundred million years of its existence, what is the probability for the emergence of life as a function of cosmic time. This very question has, in fact, been discussed by Abraham Loeb and colleagues at Oxford University, Rafael Batista and David Sloan. Using analytic expressions for the SFR, the IMF, and stellar lifetime as a function of mass, Loeb and coworkers argue that the probability of life emerging peaks at about the same time as the lifetime of the very lowest mass stars—that is, when the universe will be ~10^{13} years old.

The questions that naturally arise from this result are manifold, and not least among them, we now need to ask not only why did life appear on Earth at such an early time, when the universe was only 10 billion years old, but why is it that life evolved around a star like the Sun, which is ten times more massive than the lowest mass stars. These questions cut to the very core of the more philosophical issues that deal with the so-called Copernican principle (or the principle of mediocrity), which argues that the Sun and Solar System (and us) should be typical rather than exceptional.

Indeed, we are led by the extreme conclusions offered by Loeb et al., if correct, to reject the Copernican principle entirely and adopt a more pragmatic philosophical stance, which simply states that life exists in the universe now because the universe was primed from the very outset to produce life at a relatively early stage. There is no implied mysticism in this statement; rather it builds upon the basic understanding that whatever the processes involved in the appearance of the original spark, or sparks, that set life on its cosmic journey, they were natural processes (that is, they operated within the realm of knowable physics) that found a way to work within the framework of the physical parameters that govern the structure of the universe.

Certainly, the processes are constrained with respect to the various physical constants, but there are no specific reasons to suppose that any other sets of physical constants would mean no life. In terms of the origin of life, there really is no mystery of fine-tuning, or need to select a special universe from an infinite set of multi-universes. Rather life, as a series of chemical reactions that temporarily borrows order from natural occurring energy gradients, with the attached ability to pass on construction designs to the next generation and which are amenable to change through evolution, simply found a way to proceed within the framework of what it was given in terms of universal physical constraints. Life, the universe, and everything, to co-opt a phrase from Douglas Adams, are certainly complicated, but there is no reason to overcomplicate the philosophy of the issue, and there is absolutely no good reason to suppose that the Copernican principle tells us anything useful about how "it" all operates.

Although carbon planets might well form around CEMP stars, it is entirely possible that planets with a more Earth-like composition of silicates and metals might have formed around Population II stars. Indeed, Rosanne Di Stefano (Harvard-Smithsonian Center for Astrophysics) and Alak Ray (Tata Institute, India) argued, at the January 2016 meeting of the American Astronomical Society, that planets in orbit around the Population II stars within globular clusters may be the first places where intelligent life is identified in our galaxy.

Although only one planet has been found to date in a globular cluster,[11] Di Stefano and Ray argue that even if the planet formation rate is low, the long lifetime and the relative stability of globular cluster environments is conducive to the evolution of life. Researchers have traditionally argued that planets are not likely to survive in orbit around the closely packed stars located within a globular cluster, but Di Stefano and Ray note that should a planet form within the habitability zone of a low-mass globular cluster star, then the probability of its survival is actually quite high.

This result builds upon the fact that the habitability zone, defined as the zone around a star in which liquid water might exist on an orbiting Earth mass planet, is so close in towards the parent star that gravitational scattering by another star is unlikely to occur over the present age of the universe. Accordingly, a planet located within the habitability zone of a globular cluster star could in principle be as much as 10–12 billion years old. Given

---

[11] This is the so-called Methuselah (or Genesis) planet located in the globular cluster Messier 4. This 2.5 Jovian-mass planet (PSR B1620-26b) is actually on a circumbinary orbit about a binary pulsar and white dwarf system, and is estimated to be 12.7 billion years old, the oldest exoplanet currently known.

such a long head-start, Di Stefano and Ray suggest that globular clusters may well be the best places to look for signs of advanced extraterrestrial civilizations.

Di Stefano and Ray additionally describe what they call the "globular cluster opportunity," reasoning that since stars in a globular cluster are about 20 times closer to each other than in the solar neighborhood, then interstellar communication and travel is a real possibility. With a characteristic separation of some 10^{15} m, it would take a radio signal just 40 days to travel from one star to the next. Traveling at 10 km/s it would take 5000 years to physically travel from one star system to another; at 1/10 light speed it would take just 2 years. Given such rapid communication possibilities, Di Stefano and Ray speculate that "although individual civilizations within a cluster may follow different evolutionary paths, or even be destroyed, the cluster may always host some advanced civilization, once a small number of them have managed to jump across interstellar space. Civilizations residing in globular clusters could therefore, in a sense, be immortal."

Although the first-born stars were massive, bright, and short-lived, the last-born stars will be diminutive, faint, and almost ageless. The meek shall indeed inherit the cosmos. The characteristic time t_d required for star formation to deplete the gas content of a galaxy can be expressed as the ratio of the gas mass M_{gas} to the SFR: $t_d = M_{gas}/SFR$, where the SFR is the deduced global star formation rate. For our Milky Way Galaxy, $M_{gas} \sim 10^{10}$ $M_\odot$, and the observed SFR ~ 2 $M_\odot$/year, hence $t_d = 5 \times 10^9$ years. The actual time to deplete the gas within a galaxy will be longer than t_d, since the SFR is expected to decrease as the mass of the available gas to make stars decreases. If it is assumed that the SFR varies directly as the mass of the gas that is available to make stars, then the characteristic gas depletion time is increased by a factor of about 25 [20], giving a characteristic galactic gas depletion time of 125 billion years. This is about nine times longer than the present age of the universe.

As the gas supply in the galaxy is depleted, the formation of massive stars will begin to dwindle, and accordingly the last formed stars will be of low mass and faint. These diminutive stars are, however, very long lived, and have lifetimes measured in tens of trillions of years. Proxima Centauri [21], the closest star to the Sun at the present epoch, for example, has an estimated main sequence lifetime of 2 trillion years (some 169 times longer than the present age of the universe). These final stars will not only be of low mass, they will also grow in a metal rich environment, and this favors the generation of associated planets (recall the study by Peter Behroozi and Molly Peeples discussed at the end of Section "The Engines of Evolution").

The effect of metallicity on planetary structure and evolution is not well understood at the present time, but it is clear [22] that the radius of a pure iron planet must be smaller than that of a silicate planet having the same mass. Presumably, the iron-rich planets of the future will have larger central cores and smaller silicate mantles than those exhibited by the same-mass terrestrial planets of present times, but how this might affect important biological inducing factors such as plate tectonics, magnetic field generation, and atmospheric structure are unclear.

The long-term future of our galaxy and the consequences of the gravitational dance being played out by the local cluster of galaxies are difficult to predict, but in broad brush form it is clear that mergers of galaxies will inevitably take place. Indeed, the Milky Way Galaxy and Andromeda are due to begin coalescing in about 5 billion years from now, and within a trillion years the Local Group of galaxies will have mostly merged into one large meta-galaxy. Indeed, far-future star counts will reveal a universe of stars similar in distribution (but much larger in extent) to that envisaged by William Herschel, Jacobus Kapteyn and Harlow Shapley.

The end of star formation will occur during the formative epoch of the future meta-galaxy, and the Stelliferous era [20] will come to a close when the universe is of order 100 trillion years old. From this end-time onwards the universe will no longer be generating new stars, and the meta-galaxy will be increasingly composed of a darkling mass of degenerate black dwarfs, neutron stars, and black holes. Even at 10^{14} years of age the universe will still be a physically dynamic place [20]; it's just that there won't be any stars to illuminate the action.

For all this, however, stars have had, and will continue to have for a very long time yet to come, a bright future. The first stars appeared when the universe was perhaps just a few millions of years old, and the last will wink out when it is a few hundred of trillions of years old. By any standards of reckoning a history spanning eight-orders of magnitude in time has to be considered a very good run.

REFERENCES

1. The dynamic collapse time is the free-fall time under gravity. The gravitational force acting upon a small *blob* of material, of mass m, at the surface of a star will be $F_{grav} = G m M / R^2$, where M and R are the mass and radius of the gas cloud or star, and where G is the universal gravitational constant. The acceleration of the surface blob, a_{blob} will then, by Newton's second law of motion, be $a_{blob} = Fgrav / m = G M / R^2$. After a time t the *blobs* displacement S from its initial location will be $S = \frac{1}{2} a_{blob} t^2$. The collapse time corresponds to the time required for the blob to move through a distance corresponding to the initial radius of the gas cloud or star R. Accordingly, $\tau_{dyn} = (R^3 / G M)^{\frac{1}{2}}$. This expression is usually simplified by noting that the average density of a star is $< \rho > = M / (4/3 \pi R^3)$, and accordingly, $\tau_{dyn} = (3 / 4 \pi G < \rho >)^{\frac{1}{2}}$. If the imbalance between gravity and the pressure gradient is some fraction ε of the gravitational term, then $a_{blob} = \varepsilon Fgrav / m$ and the collapse time will be $t_{col} = \tau_{dyn} / \varepsilon^{\frac{1}{2}}$, and accordingly, for the Sun, given $t_{col} > 4.5$ billion years, so $\varepsilon < 10^{-28}$.

2. The labels red and white applied to red giants and white dwarfs are set according to Wien's law: $\lambda_{max} T = $ constant. Using the notion that the wavelength of maximum intensity λ_{max} is an approximate measure of the dominant color of a star, cool stars will have a value of λ_{max} placed towards the red end of the spectrum – hence such stars will appear red to the human eye. Likewise, hot stars will have a value of λ_{max} placed towards the shorter wavelength, blue end of the visual spectrum, and hence such stars are white to blue in color.

3. The question of solution uniqueness is still, technically, unsolved with respect to the equations of stellar structure. The so-called Vogt-Russell theorem does correctly state that once the mass and chemical composition at each point inside a star has been specified, then its internal structure is uniquely determined. What this theorem, which has never actually been proven in the general case, should really say is that a solution to the equations probably exists. As to whether there is only one solution is another issue. Part of the problem is that the four differential equations have four boundary

conditions two of which apply at the center and two of which apply at the surface. Arbitrarily integrating the equations from say the center of a stellar model outwards will not in general produce the correct surface convergence conditions. In a 1966 review of the theory of stellar structure, Thomas Cowling (University of Leeds) recalled asking Douglas Hartree (Manchester University) in 1938 if he had thought of trying to solve the equations of stellar structure on his then newly invented (mechanical) differential analyzer. Hartree apparently replied that, "we thought of having a go at the equations of stellar structure; but they are horrible equations". Indeed, the elegance and beauty of equations seen by the pure mathematician are often the horrendous nightmares of the applied mathematician. In the modern era, where rapid numerical computation is possible, it has occasionally been found that multiple solutions to the equations of stellar structure can be obtained. In general, however, it is probably fare to say that the uniqueness problem is one that the pure mathematician, rather than astrophysicist, specifically worries about. A good introduction to this topic is given in the (highly recommended) text by Rudolf Kippenhahn and Alfred Weigert, *Stellar Structure and Evolution* (Springer, 1990).

4. Payne's thesis, *Stellar Atmospheres; a Contribution to the Observational Study of High Temperature in the Reversing Layers of Stars*, was the first doctoral thesis to be awarded in astronomy by Radcliffe College (now part of Harvard University), Cambridge MA. The life story of Payne-Gaposchkin is remarkable and reveals much about the male dominated world of early 20th century academia. Payne studied botany, physics and chemistry at Cambridge University in England, but graduated without a degree since women could not receive a degree certification from the University at that time. Inspired, however, by a lecture presented by Arthur Eddington on the 1919 eclipse expedition which proved the correctness of Einstein's general relativity, Payne found a way to begin astronomical research in the new graduate program initiated by Harlow Shapley at Harvard College Observatory. Technically, while Payne did reveal in her thesis that stars must be mostly made of hydrogen, Henry Russell (the doyen of American astronomers at that time) persuaded her that the result must be wrong, and she accordingly did not push the conclusion. Payne was denied a doctoral degree from Harvard because of her gender (this is why the thesis was presented to Radcliffe College) although Otto Struve (Yerkes Observatory) was to later write that Payne had produced, "undoubtedly the most brilliant Ph.D. thesis ever written in astronomy". Ironically, Russell was eventually converted to the idea that stars must be mostly composed of hydrogen, and he is often given credit for this *discovery*.

5. Perhaps the last (heroic) hurrah with respect to solving for the equations of stellar structure with a mechanical desk-top calculator is that by Fred Hoyle and Martin Schwarzschild in 1955 – On the Evolution of Type II stars, *Astrophysical Journal Supplement Series*, **2**, 1–40 (1955). Indeed, the determination of just one stellar model would take several days worth of hand-calculation to complete. Hoyle and Schwarzschild conclude their length paper by noting that future progress in the field of stellar evolution will only be made with the, "fully automatic representation [of the equations], using large electrical machines". And this is exactly what Haselgrove and Hoyle did in a 1956 publication in the *Monthly Notices of the Royal Astronomical Society* (**116**, 515–526). It is amusing from a modern perspective to read the somewhat euphoric comment at the end of this paper that, "the machine [an EDSAC 1] is theoretically

capable of obtaining a sufficiently accurate solution to the equations in a few hours". With the introduction of the electronic computer, Louis Henyey (University of California, Berkeley) introduced in 1959 the matrix inversion technique that is almost universally used to solve the equations of structure in stellar evolution codes to this day.

6. In Jeans formula we can insert, $E_{gravity} = k\,GM^2/R$, where k is a constant related to the distribution of matter inside of a cloud of gas having mass M and radius R, and G is the universal gravitational constant, and $E_{thermal} = 3/2\,(R_g/\mu)\,T\,M$, where R_g is the gas constant, μ is the molecular weight of the gas in atomic mass units ($\mu = 2.016$ for H_2). With these two terms the collapse condition becomes $|E_{gravity}| > E_{thermal}$, and this requires $R < R_J = (2\,k\,G/3)\,(\mu/R_g)\,(M/T)$. If we assume that the gas cloud is spherical and that the gas has a uniform density ρ, then by substituting for the mass term, the collapse condition becomes: $R > [(R_g/\mu\,G)\,(T/\rho)]^{1/2}$. The collapse criterion is often expressed in terms of the Jeans mass, which from the foregoing can be written as $M_J = [(3/4\pi)(3R_g/2kG\mu)^3]^{1/2}\,(T^3/\rho)^{1/2}$. Constants aside, Jeans criterion is essentially determined by the ratio of the temperature of the molecular cloud to its density, and this makes sense if we perform a little more algebra. Indeed, given that the speed of sound c_S in a gas of temperature T is $c_S = (R_g\,T/\mu)^{1/2}$ this indicates that for collapse to begin $R \approx c_S\,t_{coll}$ – in other words collapse occurs if the time for a pressure wave to cross the cloud (R/c_S) is larger than the gravitational collapse time.

7. This follows from the scaling argument which gives the accretion rate as $M_{acc} \sim M_c/t_{coll}$, where M_c is the mass of the core growing at the center of the collapsing cloud. Given the core has density ρ_c and radius r_c, so ignoring constant terms $M_c \sim \rho_c\,r_c^3$, and since for an isothermal configuration the density varies as $\rho \sim r^{-2}$, so, to order of magnitude, the mass of the core increases linearly with radius: $M_c \sim r_c$. Now, going back to the relation for accretion, we find: $M_{acc} \sim r_c/t_{coll}$ and this must be approximately constant since the collapse time is proportional to the radius: $t_{coll} \sim r$. The argument just presented is perhaps not overly convincing, but detailed numerical simulations indicate that it essentially holds true and that the proto-stellar core, at the center of a collapsing molecular cloud, grows by accreting material at a relatively steady rate. In this manner the mass of the proto-stellar core at the center of the collapsing cloud grows at a steady rate, with the core mass at a time $t + \Delta t$ being: $M_c(t + \Delta t) = M_c(t) + \Delta t\,M_{acc}$. The typical accretion rate onto a newly formed proto-stellar core is generally taken to be of order 2×10^{-6} $M_\odot$/yr, with the in-fall velocity being about 500 m/s.

8. The angular momentum per unit mass of material is given by the quantity $L = h\,v_t$, where h is the distance from the cloud's spin-axis and v_t is the velocity perpendicular to the direction towards the cloud center. According to the angle ϕ at which the infalling material is located with respect to the spin-axis, so $h = r \sin\phi$, where r is the radial distance from the cloud center. In this manner material that is initially located close to the spin axis (where h is small and $\phi \approx 0°$) has only a small amount of angular momentum, while material further away from the spin axis (where h is of order the cloud radius R and $\phi \approx 90°$) has a large amount of angular momentum. The critical point about angular momentum is that it is a conserved quantity, and this means that for any given blob of material the angular momentum L must remain constant at all times. What this means for our collapsing cloud matter is that as it approaches the central proto-star, so the value of h becomes smaller and to compensate for this the velocity v_r must increase.

9. S. Vogt *et al*, 2007. Power Output during the Tour De France. *International Journal of Sports Medicine*, **28**(9), 756. While Lance Armstrong was the winner of the 2005 Tour, this title along with his six other Tour titles were stripped from the record books, due to doping allegations, in 2012.

10. The discrepancy between the age estimates for the Sun and Earth derived by geologists, evolutionary biologists and physicists came to a head in the late 1800s. The key antagonist was William Thomson (who was the first scientist to be awarded a peerage and is possibly better known today under his title of Lord Kelvin of Largs), who estimated in 1862 that the Sun could be no more than a million years old. To this, in a second publication in 1863, he added that the Earth had an age of order 98 million years. Kelvin's analysis was based upon classical thermodynamics. Assuming that the Earth had formed as a molten mass he calculated how long it would take for it to cool to its present temperature, repeating, in fact, a calculation and experimental result that had been presented much earlier by Isaac Newton. The logic of Kelvin's calculation and methodology was impeccable, and in spite of later criticisms he stuck dogmatically to his numbers right to the end of his life (in 1907). The problem for Kelvin was not that his calculations were wrong, far from it in fact, the problem was that his assumptions were wrong. Yes the Earth did form as a molten sphere, but the surface temperature is not governed according to a simple whole Earth cooling-off process. Likewise, Kelvin's calculations assumed that the Earth had no internal energy source – it is now known that the Earth's interior is heated by the decay of radioactive elements. As late as 1909, Mark Twain was to write in his *Letters from the Earth*, that "as Lord Kelvin is the highest authority in science now living [not actually so when the letters were published], I think we must yield to him and accept his views". This, in a nutshell, captures the problem inherent to many contentious scientific debates – authority should always be questioned and only occasionally listened to with any degree of confidence.

11. Jeans built his energy generation scheme upon ideas and physical models that are no longer held to be true. Indeed, as early as 1904 Jeans invoked the idea that energy might be liberated as a consequence of the, "annihilation of two either strains of opposite kinds". Even by 1904 this idea was somewhat outdated since the existence of an all-space-pervading ether had been ruled-out through the pivotal experiments of Albert Michelson and Edward Morley in 1887. While Jeans invoked the idea of converting matter into energy, indeed, at a time prior to Einstein's 1905 publication of special relativity in which the famous $E = mc^2$ formula appeared, the idea of antimatter was essentially solidified in its modern form in the pioneering publications of Paul Dirac in 1928. The first antimatter particle, the positron, was discovered in 1932 via cloud chamber experiment conducted by Carl Anderson – in this case the positron (the antimatter particle to the electron) was produced and observed during a cosmic ray shower event. The annihilation of electron-positron pairs, to liberate $2m_e c^2 = 1.64 \times 10^{-13}$ joules worth of energy, is in fact one of the steps in the proton-proton chain by which stars do generate internal energy via fusion reactions: the positron being produced in the first step of the process when two hydrogen atoms combine to produce a deuterium nucleus.

12. It is interesting to note that while Lockyer's meteoritic impact hypothesis was never largely popular, as late as 1927, one finds Harlow Shapley and Cecilia Payne (later and more famously, Payne-Gaposchkin) publishing the results of a study looking for the, Spectroscopic Evidence of the Fall of Meteors [sic] into Stars (*Harvard College Observatory, Circular* 317 (1927). The idea and methodology behind the study by Shapley and Payne is certainly sound, but remarkably they conclude, "the question of the alternative hypothesis that the diffuse nebulae (a) are wholly gaseous, or (b) are mainly meteoric, seems to be decided in favor of the latter". Well, of course, hind-sight for us is 20–20, and it should be pointed out that while it is now clear that stars are not powered and/or even adversely affected by meteorite, asteroid or comet impacts, such impacts will most definitely occur. The observations provided by Sun-monitoring spacecraft reveal, for example, that it is occasionally directly-struck by a wayward comet and that sungrazing comets skim through its outer corona every few months. Most cometary encounters with the Sun (and by inference other stars) end in the cometary nucleus being harmlessly destroyed, John Brown (Astronomer Royal for Scotland, and University of Glasgow) and co-workers, however, have recently looked at what happens when a particularly large cometary nucleus crashes into the Sun head-on. Writing in *The Astrophysical Journal* for 9 July, 2015, Brown along with Robert Carlson (JPL) and Mark Toner, argue that in some cases the nucleus will undergo a powerful airburst explosion and accordingly produce observable flare-like phenomena. Described by Brown as being, "supersonic snowballs from hell", even these massive cometary impacts will have no discernable affect upon the Sun's overall energy budget. Such a collision with a planet, however, is an entirely different story, as was revealed by the dramatic atmospheric *scaring* of Jupiter's upper cloud deck in the wake of the multiple comet Shoemaker-Levy 9 impacts in July of 1994.

13. The Coulomb energy between two protons a distance r apart is $U = \kappa\, e^2/r$, where κ is the Coulomb constant and e is the proton charge. The average temperature T of the gas that is required to overcome such a Coulomb barrier is $3kT/2$, where k is the Boltzmann constant. Equating these two expressions for a separation corresponding to the classical radius of the nucleus $r_{nuc} = 10^{-15}$ meters, gives $T \sim 9$ billion Kelvin. This (classical physics) gas temperature for which the constituent protons will be moving with sufficient kinetic energy that they might approach each other as close as the size of a typical atomic nucleus, is some 900 times larger than the actual temperature required for fusion reactions to begin, which is $T \sim 10^7$ Kelvin. It is entirely the effect of quantum tunneling that allows the protons to circumvent the otherwise insurmountable effect of the Coulomb barrier at lower temperature.

14. The elements deuterium, lithium, beryllium and boron are not actually produced by stars – rather, in fact they are rapidly destroyed in the interiors of stars. The origin of these elements was a mystery to B²FH and they invoked an unspecified x-process to explain their origin. It is now known that these elements along with hydrogen and helium are produced during the Big Bang, primordial nucleosynthesis – indeed, models of the very early universe are fine-tuned so as to replicated the observed abundances of these elements

15. In the modern era there is the irony that making, or explaining the existence of our universe is not considered to be the main problem issue, but rather it is explaining the existence of a universe that has all the very special features that enable the existence of observers like us. A wonderful introduction to this latter topic is given by Fred Adams (University if Michigan) in his article, *Stars in other universes: stellar structure with different fundamental constants* (arxiv.org/pdf/0807.3697v1.pdf).

16. The Eddington Luminosity is determined according to a balance being achieved between the radiation pressure force acting outward and the gravitational force acting inward. If we consider a small mass m of material at the surface of a star of mass M, radius R and Luminosity L, the inward gravitational force on the small mass will be $F_{grav} = G M m / R^2$. The outward radiation pressure force acting on the mass m, on the other hand, will be $F_{rad} = [(L / c) / 4\pi R^2] \kappa m$, where the term in square brackets is the radiation pressure, and κ is the opacity. The latter term is a measure of the ability of the stellar gas to absorb radiation, with the typical travel distance of a photon between successive absorption events being expressed as $l = 1 / \kappa\rho$, where ρ is the density. In the Sun the distance between absorption events (the so-called mean free path) is typically just a few millimeters. The dimensional units for opacity (which is usually a complex function of composition, temperature and density) are m^2/kg. A dynamic balance is achieved when $F_{rad} = F_{grav}$ and this condition yields the Eddington luminosity: $L = L_{Edd} = 4 \pi G M c / \kappa$, which is a result dependent upon the mass of the stars, but is independent of its radius.

17. The first gravitational waves from a black hole merger event were recorded by the LIGO collaboration on 14 September 2015 (recall figure 2.16). This event has been interpreted as the coalescence of two black holes with initial masses of 36 and 29 solar masses respectively. While such black hole mergers have long been anticipated, the problem now is that there is no clear pathway by which stars can actually produce black holes with the masses (apparently) required. Perhaps the best studied (stellar mass) black hole system is that of Cygnus X-1, and in this case, the black hole has an estimated mass of 15 $M_{\odot}$. A second event was detected by LIGO on 26 December 2015; this time due to the coalescence of two black holes with masses of 14 and an 8 $M_{\odot}$. The existence of black holes with masses of many tens of solar masses hints at the possibility that the upper stellar mass limit might range much higher than the presently accepted value of 150 solar masses. There is a certain irony, and indeed surprise, that the first events detected by LIGO are the result of black hole coalescences events, since it was expected that closer, that is within the Milky Way, neutron-star coalescence events would provide stronger and more common signals - as ever the Universe continues to challenge the expectations of human calculation.

18. Energy generation within Population III stars runs in a different manner to that of Population I or II stars. The latter stars contain some heavy elements, especially ^{12}C, and can accordingly generate energy via the CNO cycle (recall section 5.7). Population III stars have only hydrogen and helium within their interiors and so must initially generate energy via the proton-proton chain (PPC). Since the PPC runs at a much lower temperature than the CNO cycle it acts as a poor internal thermostat and the central core becomes inordinately dense and hot. This latter situation enables the early onset of the triple-α reaction and the star begins to form ^{12}C from helium nuclei.

The newly formed carbon then enables the onset of energy generation through the CNO cycle. With the various energy generation cycles switching on-and-off, a Population III star is highly prone to the development of pulsation cycles, and this has consequences for the final end phase. Between 10 and 100 $M_\odot$, the end phase will be that of a core-collapse supernova. For masses between 100 and ~250 $M_\odot$, the end phase will be that of a pair-instability supernova. In this latter case the temperature inside the star is so high after helium burning ends that electron-positron pairs begin to form, and this robs energy from the star and its pressure gradient can no longer support the weight of overlying layers – the star begins to collapse. This collapse increases the temperature and density of the core to such an extent that explosive oxygen burning occurs and this triggers the disruption of the entire star. For pair-instability supernovae no central core remnant survives. Black holes as an end phase of stellar evolution will only appear again once the initial mass exceeds ~ 250 $M_\odot$ and this time they will form by direct gravitational collapse.

19. See the research paper: N. Mashian and A. Loeb: *CEMP stars: possible hosts to carbon planets in the early universe* (arxiv.org/pdf/1603.06943v2.pdf). Details of the observational criteria for distinguishing between the various CEMP star types are reviewed by Jimni Yoon *et al.*, in the article, *Observational constraints on first-star nucleosynthesis. 1. Evidence for multiple progenitors of CEMP-no stars* (arxiv.org/pdf/1607.06336v1.pdf).

20. This calculation is based upon the arguments and analysis presented by Fred Adams and Gregory Laughlin in their article, *A Dying Universe: the long term fate and evolution of astrophysical objects* (arxiv.org/pdf/9701131.v1.pdf).

21. The author has previously discussed the details of this star and its companions in: *Alpha Centauri: unveiling the secrets of our nearest stellar neighbor* (Springer, New York, 2015).

22. State of the art planetary models, with a range of compositions, have been described in the research paper: *Mass-radius relationships for solid planets*, by Seager, Kuchner, Hier-Majumder and Militzer (*The Astrophysical Journal*, **669**, 1279–1297, 2007).

6

It's a Far Flung Life

I'm sure the universe is full of intelligent life.
It's just been too intelligent to come here.

—Arthur C. Clarke

The *Pillars of Creation* title given to this book is intended to emphasize the fundamental role of star formation in the transformation of an initially sterile, low-complexity universe into a dynamic and fecund environment capable of evolving ever more complex structures. Indeed, somewhere, and indeed, somehow, in the long and twisted chain of star-created opportunities that have existed over the past 12 billion years, life appeared in the universe.

It is still far from clear as to whether the origin of life, that is, the appearance of some form of metabolizing and self-replicating organic structure, was inevitable or the work of a purely fortuitous set of random and highly improbable circumstances. It seems likely that we shall never know the answer to this specific origins question, but we do know that life has evolved and thrived for many billions of years on Earth. Indeed, life appeared on Earth in what seems to be a rather rude and remarkably quick fashion.

Fossil evidence for the existence of bacterial life has been traced back to at least 3.5 billion years ago, as exemplified by the 3.4-billion-year-old Apex Chert microfossils from the ancient Pilbara craton in northwest Australia, and by the microbial activity evident in the 3.5-billion-year-old pillow lava preserved in the Barberton Greenstone Belt in South Africa [1]. Most recently Elizabeth Bell (University of California, Los Angeles) and coworkers [2] have pushed back the time at which organic life was apparently present on Earth to some 4.1 billion years ago. The evidence relating to this earliest record of terrestrial life is contained within zircon crystals extracted from the Jack Hills region of Australia, and the results bring into question all that we once thought we knew about the history of life on Earth.

Given that bacterial life was present on Earth 4.1 billion years ago, then it would appear that life was able to survive the great cataclysm of the late heavy bombardment (LHB), a time of intense and large impact basin-producing activity in the inner Solar System that took place between 3.8 and 4 billion years ago. It is usually assumed that Earth's surface must have been sterilized of life during the LHB, but the new data published by Bell and

© Springer International Publishing AG 2017
M. Beech, *The Pillars of Creation*, Springer Praxis Books, DOI 10.1007/978-3-319-48775-5_6

coworkers implies that life not only survived the LHB impacts but came into existence on a very short timescale of just 400 million years.

Given that we do not know the detailed alchemy behind the origins of primordial life, we can either accept that half a billion years is time enough for the miraculous chemistry of life to successfully accomplish its first-round of experiments,[1] or we can ask the question, did life initially evolve on Earth, or was the primordial Earth seeded by archaic bacteria that evolved somewhere else in the galaxy? This latter question is not intended to push the origins issue to some remote, even more unknowable location in space and time, but it is rather connected with the notion that the potential for life is a galactic (even universal?) phenomenon, and that our Solar System, with Earth in particular, being a highly favored site in which the explorative fingers of evolution have had the opportunity to work and generate a remarkable linage of more and more complex living entities.

Evolution by natural selection is the overriding reason for the presence of advanced life forms on Earth. Indeed, the principle of evolution by natural selection is so clearly at work on Planet Earth that is entirely reasonable to consider it as an underlying principle of astrobiology. The entire universe is evolving—physically, materially, and chemically—and these inherent changes can be selected for by an evolutionary process to favor the existence of life in almost any ambient host environment.

On Earth we find extremophile bacteria in what, at first glance, seem the most unlikely of places, from boiling-hot geothermal springs to salt-rich and high acidity ponds and even desiccated desert environments. Indeed, bacterial life is everywhere on the surface and deep within the mantle of Planet Earth. For all of the incredible diversity on life (past and present) on Earth, however, there is nothing particularly special about its basic chemical makeup. All of the atoms and molecules needed to make life work are readily found within the interstellar medium, buried deep within molecular clouds, and formed in the vast, nuclear-burning cauldrons of past generations of myriad stars.

We know where the basic material came from for life, what we do not know, at the present time, is where and when life first took its first metaphorical breath. The answer, however, seems increasingly to point towards not only an emergence of life somewhere other than on Earth but also to a time long before the Sun and Solar System even existed. This remarkable new proposition, if indeed it turns out to be correct, relies on the principle of an old idea—the notion of panspermia; literally "seeds everywhere."

Italian physicist Enrico Fermi was one of the great twentieth century pioneers of quantum mechanics and particle physics. His contributions to both theoretical and experimental physics are legendary to practitioners in the field, but to the general public he is probably best known for a paradox he never actually articulated.

---

[1] The classic laboratory experiment pertaining to the *in situ* generation of the complex molecules needed to kick-start life is that by developed by Stanley Miller and Harold Urey in the 1950s. Numerous variants of the original experiment have since been performed and it is reasonably clear that complex organic molecules can be generated within and upon the young Earth The only question now is, did the chemistry additionally ignite the spark of life *ab initio* on the raw and newly formed Earth?

Be this as it may, the Fermi (so-called) paradox lies at the heart of the modern search for extraterrestrial intelligence. The paradox, which is really just a not-unreasonable question, typically runs as follows: "If life is common within the galaxy, then why has Earth not been visited by aliens?" The question is premised on the assumption that life does exist elsewhere within the galaxy (and the greater universe), and it additionally assumes that interstellar travel is possible.

A whole cottage industry has developed around the theme of Fermi's paradox, with proponents on both sides of the argument using it to bolster their particular views. However, the paradox, or question, upon reflection, is not really well founded, and in its most commonly applied form it is not a question that is likely to have any clear or agreed upon answer. One might argue, for example, that the time to visit every region of the galaxy using interstellar spaceships that can travel at, say, half of light speed is small compared to the age of the galaxy. Therefore, the fact that Earth has not been visited by multiple advanced alien life forms must mean that no such entities exist. This, of course, is a nonsense answer to Fermi's question, and it is nonsense for so many reasons that it will suffice to introduce just two counter arguments, the first being that interstellar travel at half of light speed is an entirely unlikely possibility, even given a billion years of advanced engineering research. Second, there is absolutely no reason to suppose any alien race has any interest whatsoever in physically exploring the galaxy. The non-existence of so-called von-Neumann probes within the Solar System is an equally fallacious way of trying to answer Fermi's question.

The answer to Fermi's question, if we are to place our own cards on the table, most probably lies within the no interest in interstellar exploration domain. It is an entirely impractical and unreasonable exercise. The deep exploration of interstellar space, whether by living entities or by robots, hundreds to thousands of light years away, carries no societal or scientific value, since any data that might be gathered will never be received by the initiating party. Better, by far, to stay at home, and learn to live within one's own planetary system than to set foot into the vastness of the interstellar realm. Better by far to invest time and money into science conducted from within one's host planetary system than to send an equivalent investment of time and money into an inherently unknowable place.

Of course, this is just one of probably many hundreds of possible answers to Fermi's question. The real problem is that we presently have no solid data from which to construct any reasonable answer to the question, which is why it is a poorly framed question. However, what present-day science speculation does allow us to say, as will be argued shortly, is that the first appearance of life occurred within our galaxy about 10 billion years ago. If this is indeed the case, then the answer to Fermi's question is that alien life has very much arrived on Planet Earth, and we are it.

Although the place where life first evolved may be unknowable, the time at which life first appeared in the Milky Way Galaxy can be estimated, and such an estimate has been made by Alexei Sharov (Laboratory of Genetics, Baltimore). Presenting his initial arguments in 2006, Sharov and later Sharov and coworkers [3] have striven to discover the rate at which biological complexity has increased over time. The specific proxy that Sharov identifies as a measure of biological complexity is that of the non-redundant functional genome size. Essentially, the more complex a species is, the larger the genome size required to code for that complexity must be.

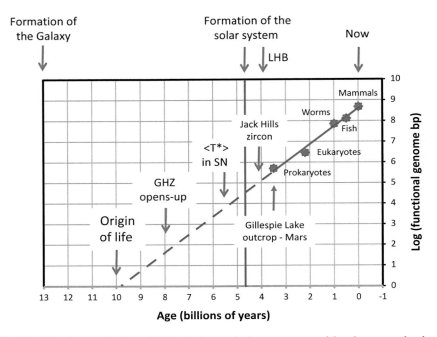

Fig. 6.1 The time evolution of biological complexity, as measured by the non-redundant functional genome size (see text for details). The functional genome data points are from Sharov [3]

For modern-day mammals the non-redundant functional genome size is of order 480 million base pairs; for the more ancient and less complex prokaryotic bacteria the number is some 500,000 base pairs. Bringing together first fossil appearance times and the non-redundant functional genome estimates, a graph (see Fig. 6.1) of biological complexity against time can be constructed. Incredibly, as seen in Fig. 6.1, it would appear that biological complexity has increased as a power law in time with the genome size increasing by a factor of about 10 for every 1 billion years of time.

That biological complexity has increased over time should come as no surprise. Indeed, evolution by natural selection essentially tells us that this result is to be expected in the continually changing environments found on Planet Earth. What is perhaps more intriguing about the graph in Fig. 6.1 is that extrapolating the data line backwards to the time at which Earth and the Solar System first came into existence, the non-redundant functional genome indicator is not zero. The extrapolated data line, in fact, does not cut through the time axes until a point set some 10 billion years into the past. This intercept point, assuming the power law development of complexity has always held true, corresponds to the moment of creation—the time at which the first, simplest possible life forms appeared. Remarkably, the evolution of biological complexity on Earth suggests that life first appeared in our galaxy some 5 billion years before Earth and our Solar System even existed, and within 3 billion years of the formation of the very first stars.

Figure 6.1 is a little crowded, but it includes a number of important cosmic and Solar System time markers. The solid vertical line at 4.56 billion years ago indicates the time, as deduced by the study of meteorite ages, at which the Solar System formed. The Late Heavy Bombardment marker is set at 3.9 billion years ago, and close to this at 4.1 billion years in the past is the formation time of the Jack Hills biogenic carbon containing zircon.

Following the detailed studies conducted by Charles Lineweaver (University of New South Wales, Australia) and coworkers, the approximate time at which the galactic habitability zone (GHZ), located between 7 and 9 kpc from the galactic center, first opened up is set at 8 billion years ago. The average age of stars <T*> in the region of our solar neighborhood (SN) is further indicated, and set on the basis that the majority of stars within the SN are at least a billion years older than the Sun. The data point for the emergence of the simplest prokaryote bacterium is set at some 3.5 billion years ago, consistent with the age of the Apex Chert microfossils found in Australia and with the microbe-activity bearing pillow lavas preserved in South Africa. To these data points we have also added a marker to show the 3.7-billion-year-old Gillespie Lake sandstone outcrop located on Mars and investigated by NASA's Curiosity rover.

This latter marker is added on the basis of Nora Noffke's (Old Dominion University, Norfolk, VA) detailed study [4] of the preserved structures (so-called, microbially induced sedimentary features, or MISS) in what was evidently a shallow lakebed region. Indeed, numerous features, interpreted as being fossilized microbial mats, are evident in the Gillespie Lake region. Intriguingly, it is possible that bacterial life may have existed on Earth and quite probably on Mars within 100,000 years of the Solar System coming into existence.

The off-Earth origin for life provides one possible explanation for the very rapid (that is, within a few hundred million years) appearance of microfossils in the oldest known rocks on Earth. With this idea, however, we have still not solved the problem of how life first began, but the implications are that it started somewhere else within the galaxy, and that the essential bacterial spores and/or the vectors encoding information for elementary life are prevalent within the interstellar medium. Not only are the ingredients for life plentiful in the Solar System and the interstellar medium, but so, too, it would appear, are the bacterial seeds enabling primitive life to evolve given the right set of circumstances, that is, the formation of a terrestrial planet with an associated atmosphere located within the habitability zone of some newly formed star.

That our Solar System still contains reservoirs that host the basic chemical ingredients necessary for the evolution of life is evidenced by studies of carbonaceous chondrite meteorites and cometary nuclei. Indeed, instruments aboard ESA's Rosetta spacecraft, in orbit around comet 67P/Churyumov-Gerasimenko (Fig. 6.2), have recorded spectral signatures indicating the presence of the amino acid glycine and of phosphorus, the latter being a key component of DNA and cell membranes. Comets such as 67P/Churyumov-Gerasimenko have hardly changed their basic chemical makeup since the time the Solar System formed, 4.5 billion years ago, and comets have additionally collided with Earth since it first formed, discharging, at least initially, their vital chemical stash into the (pre)biotic soup that eventually nurtured the appearance of complex life. In 1979 progressive rock musician Frank Zapper penned the lyric, "It makes its own sauce … if you add water," and it would seem that the same mantra might be applicable with respect to the evolution and seeding of life within the greater galaxy.

Fig. 6.2   The nucleus of comet 67P/Churyumov-Gerasimenko. In May of 2106, ESA scientists announced the detection of glycine and phosphorus within the comet's coma. Image courtesy of ESA

The idea of panspermia, and the seeding of life on a planet via an external agent, has a long history. And, although it was certainly a philosophical idea debated by astronomers in the mid to late nineteenth century, it was Swedish chemist Svante Arrhenius who, in 1908, first developed a detailed outline of how the process might actually work. Arrhenius essentially introduced the idea of starlight-riding bacterial spores spreading through interstellar space and on occasion becoming bound up within star-forming regions. On this basis, the essential encoding for life was directly incorporated into any planet forming gas and dust disks, and as soon as the conditions on a planetary surface were amenable to supporting it, life could establish its first toe-hold and thrive.

Some evidence for at least the presence of microscopic carbon-rich material in interstellar space was found by British mathematician, physicist, and astrobiologist Chandra Wickramasinghe (Buckingham Centre for Astrobiology) in the early 1980s. Working in collaboration with the late Sir Fred Hoyle (University of Cardiff), Wickramasinghe posits that several distinct absorption features, in the wavelength range between 3 and 4 μm, in the spectrum of numerous infrared sources (interstellar dust clouds and dust-enshrouded stars) are due to the presence of organic carbon. Although, as we have seen in Chap. 3, interstellar absorption is generally attributed to the scattering of starlight by interstellar silicate and graphite grains, Hoyle and Wickramasinghe specifically argued in 1980 that organic carbon, as revealed by the observed absorption line features, has a biological origin [5].

Indeed, Hoyle and Wickramasinghe suggest that between 25 and 30 % of the interstellar medium could be in the form of bacteria-like grains. This latter possibility is still contentious, and if true it provides strong support for the panspermia paradigm. However, a biological origin for the carbon grains has yet to be established.

Many versions of the panspermia idea have been developed since the time of Arrhenius, and in recent times the specific notion of lithopanspermia has found considerable favor among a growing number of researchers. The idea of lithopanspermia was first outlined in some detail by William Thomson (1st Baron, Lord Kelvin of Largs) in his 1894 presidential address to members of the British Association for the Advancement of Science. Here the idea is that rather than the bacterial spores being free-flying (and therefore continuously exposed to cell-destroying UV radiation and cosmic rays), they are embedded within and on the surface of sheltering silicate dust grains. Indeed, local to the Solar System, it has been suggested that bacterial life may have first evolved on Mars and was subsequently delivered to Earth through meteorite transportation—a journey that we know is possible, since Martian meteorites have been collected on Earth in the modern era.

Brett Gladman (University of British Columbia, Canada) and coworkers have further modeled the orbital evolution of material ejected (via asteroid impacts) from Earth's surface, and find that on timescales of order 5 million years such material can spread throughout the entire Solar System [6], and potentially, thereby, seed the ice-rich and subterranean sea-supporting Jovian moons such as Europa, Titan, and Enceladus. William Napier (Armagh Observatory, Ireland) has additionally studied [7] the fate of meter-sized boulders ejected from Earth, and while it would seem that they are substantially reduced in size through collisions with zodiacal cloud dust particles, small fragments may nonetheless safely end up leaving the Solar System for interstellar space. Napier additionally suggests that, assuming the ejection of bacteria bearing rock is common in other planetary systems, the entire galactic disk could be seeded within a few billion years.

Although the exchange of meteoritic material between the planets and their moons is entirely possible within our Solar System, it has recently been suggested [8] by mathematician and artist Edward Belbruno (http://www.ebdelbruno.com) along with coworkers from Princeton University, University of Arizona, Livermore National Laboratory, and the Centro de Astrobiologia in Spain, that bacteria-bearing material may have additionally been shared between a number of the stars within the Sun's birth cluster. Belbruno et al. specifically develop a lithopanspermia model based on the idea of low velocity, weak transfer orbits in which material that is only weakly bound gravitationally to one star can migrate towards and be captured by another nearby companion. This weak transfer process could, Belbruno and coworkers argue, have operated within the Sun's birth cluster for an estimated time of several hundred million years, allowing thereby plenty of time for the exchange of material between the planets within our Solar System and any exoplanets that might have formed in orbit around the Suns stellar siblings.

Bacteria were the first living creatures to inherit the Earth and they are the hardiest of entities, but can bacteria survive in space? The answer to this is yes, and this answer highlights one of the important points about the panspermia hypothesis as a means of spreading life in the galaxy—it is testable. The process of panspermia makes very specific predictions and requires certain properties of bacteria. In principle the predictions and requirements are testable either in the laboratory or by direct spacecraft exploration of the Solar System. The panspermia hypothesis, for example, makes the straightforward prediction that any bacterial lifeforms that might be discovered, for example, on Mars, Europa, or Enceladus, which are locations suspected of either once and/or currently harboring life, should be biologically similar to the bacterial life forms presently found on Earth—that is, they should use the same DNA and RNA mechanisms for information transmission and reproduction.

However, if truly alien (sometimes called shadow life, in the sense of having no similarity to terrestrial life) bacteria are found in the available habitable locations of the Solar System, then this will immediately negate, or at the very least severely weaken, the panspermia hypothesis. The panspermia hypothesis additionally requires that bacteria be able to survive extremely long periods of dormancy, and this requirement seems to be well attested to. Indeed, degraded but viable bacterial spores have been found within ancient ice cores, with ages between 100,000 and 8 million years, extracted from the Antarctica ice field. Viable bacterial spores have also been extracted from the gut of a bee preserved in amber for the last 25–40 million years. Furthermore, bacteria carried into Earth orbit aboard the European Space Agency's EXPOSE experiment have survived, when appropriately shielded against UV radiation, and multiple years of exposure to the extreme cold of the space environment [9]. Bacteria are hardy beasts indeed.

Not only must bacteria be able to survive long bouts of deep freeze in high radiation environments, they must also be able to withstand the extreme conditions of atmospheric entry or direct ground impact. This again is a property of bacteria that can be directly experimented with. The European Space Agency's BIOPAN experiment, for example, has subjected microorganisms to short bouts of space exposure, weightlessness, and the extreme heat of spacecraft re-entry [9]—and, once again, while often severely depleted in numbers some of the bacterial specimens can and do survive. Not only do bacteria survive under artificial experimental conditions, it has long been known that meteorites can deliver complex organic molecules to Earth's surface. The Murchison carbonaceous chondrite meteorite, for example, which fell in southeastern Australia on September 28, 1969, was found to contain, among numerous complex molecules, the common amino acids glycine, alanine, and glutamic acid.

More contentiously in recent times, a number of research groups have argued that signs of biotic alteration are present in a number of different Martian meteorites. A detailed, high resolution imaging study of the Nakhla meteorite,[2] which fell in Egypt on June 28, 1911, for example, shows the same numerous dendritic track signatures of microbial intrusion (Fig. 6.3) as recorded in the 3.5-billion-year-old pillow lavas found in the Barberton Greenstone belt region of South Africa [1]. Researchers at the University of Manchester

---

[2] Part of astronomical folklore has it that one of the Nakhla meteorite fragments struck and killed a dog; this is now known to be an entirely apocryphal story.

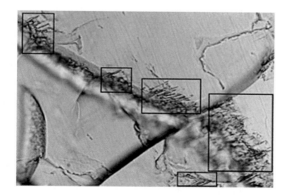

Fig. 6.3 Electron microscope image of a section of the Nakhla meteorite showing evidence of numerous thin tracks, possibly caused by bacterial invasion. Analysis of the meteorite's mineralogy indicate that it formed during volcanic eruptions on the surface of Mars some 1.3 billion years ago and that it was exposed to water for at least 600 million years prior to its ejection into space about 11 million years ago. In 2014 researchers found similar microbe-produced tracks in another Nakhla-type meteorite, Yamato 593, which was found in the Yamato Glacier region of Antarctica in 2000. Analysis of the Nakhla sample by Martin Fisk and coworkers was reported in the March 21, 2006, issue of the journal *Astrobiology* [10]. Image courtesy of Oregon State University

found further signs of a potential biotic signature in a sample of the Nakhla meteorite in 2014, with scanning electron microscope imaging revealing a strange clay-ovoid with features that hinted at a possible biotic origin.

The direct observation of fossilized nanobacteria within the Martian meteorite ALH84001 has been described by David McKay in several research papers published between 1996 and 2009, and though many research groups have argued against McKay's and coworkers' interpretations, the possibility that fossilized Martian bacteria, carried to Earth on a fragment of basaltic rock that crystallized on the surface of Mars some 4.1 billion years ago, has been detected, remains a viable hypothesis [10]. If true, and this continues to remain a contentious point, what an incredible story it is that the fossilized nanobacteria and the biotic invasion tracks within Martian meteorites tell us.

In the case of lithopanspermia it is not only the long-term survivability of bacteria in the space environment that needs to be considered, but so, too, must the bacteria-carrying substrate. The issue at play with respect to the rock substrate is that of space weathering and erosion via impacts with interstellar dust grains. Rock substrate survivability lifetimes have recently been calculated by Mauri Valtornen (Tuorla Observatory, Finland) and coworkers [11], and they find that a 1-m sized boulder might provide its residential bacteria with protective shielding for some 25-million years and travel a characteristic distance of about 15 pc (about 50 light years); a 5-m diameter boulder will survive much longer and provide any host bacteria with protection for some 450-million years and carry its resident bacteria a characteristic distance of about 225 pc (some 730 light years). Such characteristic lifetimes and carrying distances could certainly result in the extensive mixing of bacterial material within a typical star-forming region and more specifically the Sun's birth cluster.

One of the (many) vexing problems with the Fermi paradox (or question, as described earlier) is that it is essentially impossible to develop reasonable expectations from just one example. At the present time the only known location in the entire universe where life thrives is on Planet Earth. There is no other positive example of life existing anywhere else within the galaxy, at any time in the past, once we move beyond the boundary of Earth's lower atmosphere. This one data point can potentially mean everything or nothing. The question now is not that of the prevalence of life within the galaxy, which at a microbial level seems all but assured, but the emergence of intelligent and articulate life—that is, have other civilizations emerged within the galaxy that are either interested in signaling (directly or passively) their existence to anyone who might be listening in and/or watching. Such extraterrestrial civilizations might be detected via what are now traditional optical or radio telescope programs, such as SETI, or they might be revealed serendipitously, expressed as oddities, within extensive optical and infrared wavelength sky survey data collected for other reasons than SETI.

The "WOW" signal [12], recorded on August 15, 1977, and encoded within the Ohio State University Big Ears radio telescope signal intensity output data as 6EQUJ5, is one of the most enigmatic of possible ET detections within the history of the search for extraterrestrials using radio telescopes (Fig. 6.4). The signal, which lasted for some 72 s, was a one-off event, and has apparently never been repeated—and this in spite of many follow-on observations both at Ohio State and with other radio telescopes.

Although challenging, it is more than likely that the WOW signal does not indicate the detection of an actual extraterrestrial beacon. As Paul Davies (Arizona State University) articulated in his book *The Eerie Silence: Renewing our Search for Alien Intelligence*

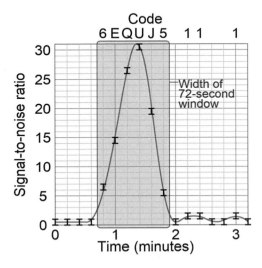

Fig. 6.4 The intensity versus time plot of the WOW signal, which lasted for 72 s. This is the expected time interval over which an extraterrestrial signal would be carried through the antenna aperture by Earth's rotation. A second antenna system should have picked up the same signal 3 minutes after the first detection; no second detection was observed

(Mariner Books, Boston and New York, 2010), after more than 50 years of radio SETI we appear, as far as can be told, to be engulfed within a rather unsettling silence. There are many possible reasons for this apparent lack of cosmic chatter. Perhaps there are absolutely no extraterrestrial intelligent life forms anywhere else in the galaxy (or the universe)—we being, therefore, by default, the proverbial "it." Perhaps we are not looking or listening in the right way. Perhaps, as articulated by physicist Stephen Hawking, intelligent life carefully hides itself away and minimizes its chances of outsider detection. Perhaps we have already detected signals from numerous extraterrestrial civilizations but haven't realized, or more specifically recognized, it yet.

Indeed, Derek Pugsley has nicely articulated this latter point in the *Journal of the British Interplanetary Society* (Volume **61**, 20–23, 2008) that, "the real problem for SETI is not **how** to search for extraterrestrial intelligence but whether humans have the capability to recognize intelligence in an extraterrestrial." Perhaps advanced (so-called) civilizations don't survive for very long. Indeed, as recently argued by Adam Frank (University of Rochester) and Woodruff Sullivan (University of Washington), predicting the temporal trajectory of any civilization is a formidable problem to deal with, and one that requires knowledge of such symbiotic variables as population dynamics, planetary carrying capacity, climate forcing and required energy supply [13].

Perhaps advanced technological civilizations end suddenly, destroying themselves through the militaristic use of the very technology that brought about their industrial advancement. Even such end-of-civilization events might be detectable if one is looking in the right place at exactly the right moment. The observational detection of nuclear detonation signatures (bright, short-duration, and correlated optical and gamma-ray flashes), or the realization that the atmosphere of a previously known exoplanet has suddenly been forced into a nuclear winter phase, could be taken as evidence for the catastrophic demise of an advanced (?) civilization [13]. Indeed, philosopher Nick Bostrom (Future of Humanity Institute, Oxford University) has forcibly argued that "The most serious existential risks for humanity in the 21st century are of our own making. More specifically, they are related to anticipated technological developments." Bostrom has further argued that the Fermi paradox (question) may be explained in terms of a "Great Filter." The filter not only acts against the development of advanced life in the first place, it also acts against the emergence of any intelligent, self-aware, tool-using, language-developing, long-enduring species.

Although radio SETI may yet succeed in its hoped for goal, alternative search methods are beginning to come into their own, with at least some of the more extreme SETI possibilities being put to experimental test. From the available survey data, for example, there is no current evidence to support the existence of any Kardashev Type III civilizations within the observable universe. Such civilizations, as first outlined in the energy-requirement scheme developed by Nicolai Kardashev in the 1960s, would required the entire energy output of their host galaxy to provide for their needs.

Key to identifying a Kardashev Type III civilization, it is assumed, would be the detection of a galaxy that is relatively dim at optical wavelengths but relatively bright in the mid-infrared part of the spectrum. To this end, Roger Griffith (Penn State University) and coworkers [14] have analyzed the infrared survey dataset of some 100,000 galaxies observed by the Wide-field Infrared Survey Explorer (WISE) spacecraft. In total, the study revealed a subset of 98 galaxies with unusual infrared signatures and a smaller set of 5 additional galaxies that were singled out as being worthy of more detailed investigation. A follow-on study by Michael Garrett (Netherlands Institute for Radio Astronomy), however, found that none of the sample set found by Griffith and coworkers were truly anomalous (in the sense of not being describable in terms of known and natural astrophysical processes). Garrett concludes that, "Kardashev Type III civilizations are either very rare or do not exist in the local universe." [14] Of course, another way of interpreting Garrett's conclusion is that Kardashev Type III civilizations, if they exist, don't display the signatures that we have assumed they should.

Planetary and host-star energy level, so-called Type I and Type II, civilizations may exist within the galaxy, but they currently remain well hidden from our gaze. Indeed, the classic and highly compelling Dyson sphere scenario, first articulated by Freeman Dyson in 1960, in which an advanced civilization enshrouds its host star within a network of energy capturing satellites, industrial platforms, and habitation colonies, has found no support from the present-day astronomical survey data (although see below). Indeed, expected to stand out as conspicuous infrared objects, Dick Carrigon (Fermilab program, Batavia, Illinois) has looked for potential Dyson sphere signatures within the data catalog developed by the Infrared Astronomical Survey (IRAS) telescope. From a dataset of some 11,000 objects with low resolution spectra, 16 were identified as "somewhat interesting" and 3 as "most interesting." Figure 6.5 shows the spectrum of IRAS 20369+5131, one of the "most interesting" objects [15]. The sensitivity of the IRAS telescope, Carrigon notes, is such that a Dyson sphere with an energy output equal to that of the Sun would be detectable out to a distance of some 300 pc. In the volume of space within 300 pc of the Sun there are of order 11 million stars, of which about 900,000 will be Sun-like stars.

A not unreasonable search strategy to adopt, as an alternative to the radio searching out of advanced galactic civilizations, is to look for signs of advanced space propulsion. As a means of travel or exploration, interstellar travel only makes sense if it can be done at extremely high speeds—indeed, at speeds comparable to that of light, the maximum speed

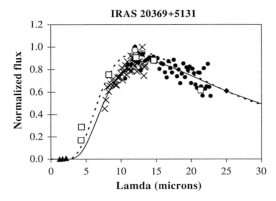

Fig. 6.5 Energy flux spectrum of IRAS 20369+5131. The closest spectrum to that expected of a Dyson sphere found in the IRAS data survey [15]. The data points are consistent with a near pure blackbody radiator with a temperature of 376 K. Assuming the source has the same luminosity as the Sun, the distance to IRAS 20369+5131 is some 42 pc (137 light years). Image adapted from ref. [15]

allowed under the constraints of general relativity. To achieve such travel speeds engines of an extra-special design will be required, and it is the signature of the propulsion mechanism that might be used to betray at least the technology of an advanced civilization.

Renowned space exploration advocate Robert Zubin, as long ago as 1995, discussed the idea of detecting extraterrestrial activity through the copious emission of gamma rays, a byproduct emission from a spacecraft using any form of antimatter drive. Additionally, Ulvi Yurtsever and Steven Wilkinson, both researchers at the Raytheon Company, have pointed out in a 2015 publication that any spacecraft traveling close to the speed of light will strongly interact with photons associated with the cosmic microwave background and thereby produce what should be a detectable gamma ray signal [16]. While astronomical satellite surveys have now pinpointed numerous gamma ray sources, none shows any signs of the motion that would be expected to accompany an interstellar spaceship.

During his historic speech that launched the beginning of the Apollo Moon landing program, presented at Rice University in Texas on September 12, 1962, U. S. President John F. Kennedy remarked that, "We choose to go to the Moon in this decade and do the other things,[3] not because they are easy but because they are hard, because that goal will serve to organize and measure the best of our energies and skills." The same essential ethos applies to SETI. The search for extraterrestrial life or intelligence is a very difficult problem, and it is not entered into on the basis that results will be easily gained. The chances of detecting another planetary civilization are always going to be small, but we assume (admittedly only armed with speculation and essentially no hard scientific justification) that they are not zero.

Indeed, the answer to the even simpler question than that posed by Enrico Fermi, that is, where are they, remains unanswered, and we are presently left with the possibility that we are either the first kids on the block or that we have not, as yet, found the right way of looking for or recognizing our advanced interstellar neighbors. Indeed, continuing the rallying theme of political posture and bravado, U. S. 21st Secretary of Defense Donald Rumsfeld became (in)famous for his "unknown, unknowns" statement with regard to the existence (or not, as the case turned out) of weapons of mass destruction in Iraq at the time before the second Gulf War. Rumsfeld, however, is entirely correct in the statement that we are indeed surrounded by unknown unknowns. It is the classic case of Plato's *Meno*, written circa 385 B. C., in which an exasperated Meno (the handsome youth from Thessaly) expounds that if you don't know what you are looking for, how will you recognize it when you come across it.

The answer to the Fermi question, and the means by which extraterrestrial civilizations might eventually be identified, most likely reside somewhere among our present-day unknown unknowns. Standard radio and optical techniques may yet prevail, but it is perhaps just as likely that we will stumble across the first extraterrestrial civilization in some entirely unexpected way and in some entirely unlikely place. Remarkably, as this book is being written, such a set of circumstance may have come about. The object in question is the star KIC8462852, one of the objects in the Kepler Input Catalog.

The Kepler spacecraft was launched by NASA in 2009, and for 4 years stared myopically at the objects within a fixed 115 square degree region of the sky in the direction of the constellations Cygnus and Lyrae. The mission was designed to monitor, in a near

---

[3] These were the development of technologies for the greater good of humankind.

continuous fashion, the brightness of more than 145,000 stars, with the intention of detecting the slight, but periodic, dip in brightness of those stars supporting exoplanet systems that produce transits in the detector system's line of sight. The Kepler mission has been a tremendous success, and as of November 2015 some 1031 confirmed exoplanets associated with 453 stellar systems have been identified, with an additional 4696 planetary candidate systems awaiting further analysis.

The modeling of the photometric variations, due specifically to planetary transits, is a well understood field of astronomy, and much information about planetary structure as well as planetary orbits can be extracted from the Kepler data. In addition the data can be used to investigate properties of the parent star, revealing the presence of binary companions, intrinsic variability, sunspot activity, and rotation. It was surprising, therefore, when a data stream defying all standard explanations appeared, and this is the story with respect to KIC 8462852.

The unique and entirely unexpected light curve displayed by KIC 8462852 were first noticed by Planet Hunter volunteers viewing Kepler derived data by eye under the zooniverse citizen science network. The Kepler spacecraft recorded photometric data on KIC 8462852 every 30 minutes, for a total observing time of 4 years, and the resultant energy flux versus time data is shown in Fig. 6.6. Within the first 150 days of data gathering the star was identified as an object of interest, since a slight, smaller than 1 %, dip in brightness was noted near day 130, this being the typical signature expected of a companion planet. A second small brightness dip was recorded near day 260, suggestive that KIC 8462852 was the host to an exoplanet with an orbital period of about 130 days.

The typical scenario after such detections would be to refine the exoplanet orbit, looking for additional small brightness variations around days 390, 520, and so on. Strangely, no such brightness variations were recorded, and the light curve remained stubbornly flat. Remarkably, however, between days 788 and 795 the light curve first began to ripple and then dive to a massive transit depth of some 15 % light blockage. This was unprecedented, and no planet, not even a Jupiter-sized planet on an orbit very close to its parent star, could produce such an effect. Then followed 400 days of nothing; the light curve remained steadfastly constant and unvarying.

A small dip was recorded close to day 1200, with another subsequent 300 days of no variation being observed. Starting around day 1500, however, all kinds of crazy phenomena were observed, with the system showing spectacular dips in brightness (some as deep as 20 % transit obscuration) for about 60 days. The latter brightness variations additionally showed extremely complex, short timescale variability, producing a light curve that no single or multiple set of exoplanets could possibly generate.

Following worsening problems with the Kepler spacecraft pointing control, no more data on KIC 8462852 was gathered after day 1580. The proverbial cat, however, was now out of the bag, and KIC 8462852 was a confirmed anomaly—not only an anomaly, however. It was unique. No other star in the Kepler archive has (at least so far, and recall there are still many thousands of systems to investigate) a light curve anything like KIC 8462852. The obvious question now, of course, is what is going on with respect to this strange system.

The remarkable properties of KIC 8462852 were first brought to public attention in early September 2015 by way of a preprint submitted to the arXiv web server (hosted by Cornell University), with the paper's lead author being Tabetha Boyajian (Yale University)

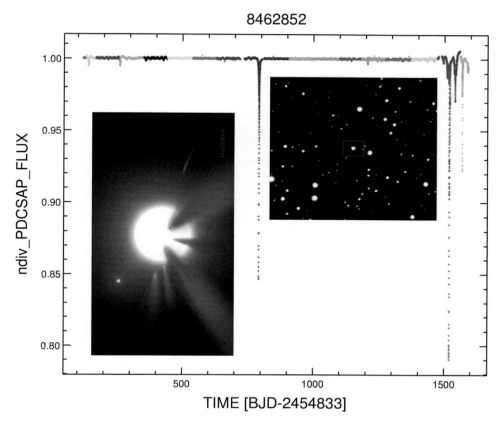

Fig. 6.6 The normalized light curve for KIC 8462852 showing the clear presence of numerous highly variable and apparently aperiodic transit events. The *leftmost inset* image shows an artist's impression of a swarm of comets in orbit about KIC 8462852—a scenario that possibly accounts for its strange light curve. The *rightmost inset* is an infrared image of the star field containing KIC 8462852 (located at image center). Data and diagram from [17] and inset images courtesy of NASA

and 28 co-authors [18]. Submitted to and eventually published in the April 21, 2016, issue of the prestigious astronomy journal *Monthly Notices of the Royal Astronomical Society*, the research paper by Boyajian et al. caused an immediate media storm.

The media frenzy was not generated because of any specific claim made by Boyajian and her coworkers, but by the fact that the team failed to develop a truly convincing explanation for the light curve variations. Here, we are being somewhat unfair to the authors of the paper. Boyajian and coworkers had developed a coherent and detailed model in which the transients observed in the light curve were due to the break up event of a large comet-like body. The strong transients near day 800 and day 1500 were, thereby, the result of a large and extended debris cloud passing in front of the star (as seen by the Kepler spacecraft). The smaller transients, presumably, were due to lesser sized fragments or dust clouds.

The implied orbital period of the dust cloud is some 750 days, and this suggests more deep and complex transients should be observable in April 2015 and May 2017. The first transit date has now passed, and no follow-on observations of the system were made (a missed opportunity, for sure). Astronomers will hopefully be motivated and ready to observe the system at the time of the predicted 2017 transit. Future observations of KIC 8462852 will be of vital importance to determine just what is going on. The dust cloud model developed by Boyajian et al. can be made to work with respect to the available data, but it is not as yet a fully compelling model. The odds of actually observing such a break-up or collision event around any individual star are remarkable small (but certainly not zero), and as Boyajian and co-workers concede in the conclusion of their paper, the infrared flux enhancement that would be expected in the wake of a collision/break-up event are not apparent.

Well, this is not the first time some complex astronomical phenomenon has gone unexplained, so why the fuss about KIC 8462852? The media fuss, it turns out, was due to the possibility, articulated by several researchers working independently of Boyajian's group, that the light curve variations observed for KIC 8462852 are exactly what one might expect if an alien civilization was constructing a Dyson sphere-like structure around its parent star. The question was forced upon the astronomical community: Is this the first detection of an alien civilization?

The way in which many astronomers set about answering media questions concerning KIC 8462852 was very revealing, and clearly identified a conservative body of researchers not only not prepared to speculate but entirely unprepared to conceive of the idea that an alien civilization might actually have been found. This situation is to some extent exactly as it should be. Scientists must, by the very nature of their occupation, be conservative in their approach to new problems, with well attested facts preceding any specific conclusions. On the other side of the coin, however, some commentators *de facto* rejected the possibility that KIC 8462852 might host an alien civilization. It was as if the very idea that something so incredibly unlikely had actually been observed was scientific anathema.

At this stage it should be said that an explanation of the light curve variation observed with respect to KIC 8462852 may well have an entirely natural explanation, the variations may yet be explained by exactly the model presented by Boyajian et al. But, the lesson that we have learned here, so far, is that if humanity is to ever achieve a level of understanding from which an alien civilization might be recognized, then the idea should be more routinely invoked as an acceptable and indeed viable explanation when standard scientific models apparently fail. The alien option, for sure, is always going to be a long shot, and it certainly would have been a wrong interpretation of the data pertaining to the discovery of the first pulsar, or the first detection of gamma-ray bursts. But, to reiterate an earlier point, we will most probably find alien civilizations in those places that are presently considered as unlikely locations.

In terms of the present-day astrobiological paradigm, KIC 8462852 would not be the kind of star expected to host an advanced extraterrestrial civilization. The reasons for this are various, but not obviously fatal to the emergence of life. Indeed, KIC 8462852 is classified as a F3 IV/V spectral-type star, and this makes it hotter, more luminous, and more massive than the Sun [19]. None of these factors, on their own, negate the possibility of habitable planets forming, but the higher temperature dictates a greater flux of ultraviolet radiation in the habitability zone—more UV radiation, for example, than the Earth receives from the Sun—and this radiation, as we have seen earlier, will work against the survivability of even bacterial life.

The luminosity class IV/V designation for KIC 8462852 indicates that it is relatively young star, and it is probably no more than a few hundred million to perhaps a billion years old. Indeed, deriving a good age estimate will probably be of vital importance with respect to interpreting the KIC 8462852 system. Unfortunately, the age is one of the least well constrained features of the star. Boyajian et al. derive a mass and luminosity of 1.43 $M_\odot$ and 4.68 $L_\odot$ for KIC 8462852, but note that the star's space motion and rotation rate fall outside of the range where good dynamic age or gyrochronology estimates can be made. The space motion of KIC 8462852 suggests that it belongs to the young disk population of the galaxy, and this fixes the ~1 billion year upper age limit indicated above.

If KIC 8462852 harbors an advanced, Dyson sphere megastructure building civilization, therefore, then life must not only have appeared very, very quickly within the system, it must have also advanced extremely (if not embarrassingly) rapidly. On Earth it took nearly 4 billion years of evolution before a dominant, tool-using, spacefaring, and intelligent animal evolved (presumptuously taking ourselves to be the peak of intelligence thus far attained on Earth). Additionally, the main sequence lifetime of KIC 8462852, in which it can generate energy through the fusion conversion of hydrogen into helium and prior to its more luminous giant phase, is estimated to be of order 4 billion years,[4] which is less than half the main sequence lifetime of the Sun.

Not only does it appear that KIC 8462852 is a young star (the Sun and Solar System being potentially 3.5 billion years older), it is also a star that will not have a viable habitability zone for very long. Intriguingly, at the present time, however, an object (debris cloud) with an orbit period of 750 days will be placed some 1.6 AU from KIC 8462852, and this, given the star's luminosity, places it right on the inner edge of the present habitability zone. Indeed, Satoko Sato (University of Texas at Arlington) and coworkers estimate that the zone over which a star such as KIC 8462852 might nurture and allow liquid water to exist on an Earth-like planet falls in the orbital range 1.6–3.3 AU [19].

As a target of special opportunity the Allen Telescope Array (ATA), operated by the SETI Institute in Mountain View, California, has begun monitoring the region of KIC 8462852 for unnatural radio signals, and these studies will be complemented in the visual part of the electromagnetic spectrum by astronomers at the Boquete Optical SETI Observatory in Panama, these later observations attempting to catch any telltale bursts of powerful laser communication emissions. On announcing the new initiative Senior Astronomer Seth Shostak commented, "On the basis of historical precedent, it's most likely that the dimming of KIC 8462852 is due to natural causes. But in the search for extraterrestrial intelligence, any suggestive clues should, of course, be further investigated – and that is what the SETI Institute is now doing." Shortly after the initial announcement was made, the first results from the radio SETI survey were published online on November 5, 2015 [20]. Between the 15th and 30th of October (2015) the ATA conducted a survey of the region surrounding KIC 8462852 in the microwave frequency region between 1 and 10 GHz. No "persistent technology-related signals" were detected to a radio transmitter power limit of 10^{20} W.[5] Not only do the first ATA results indicate a

---

[4] This is based on the main sequence lifetime varying as $t_{ms}(\text{Gyr}) = 10 \, (M_{Sun}/M_{star})^{2.5}$, and the mass of KIC 8462852 is taken as 1.43 times that of the Sun [19].

[5] This can be contrasted against Earth's strongest transmitter power of some 10^{13} W.

continuation of the long, eerie silence, they also argue against the ubiquitous use of microwave-driven spacecraft in servicing the supposed mega structure [20].

Following a few months hiatus from the Internet news and press-release websites, KIC 8462852 resurfaced as an object of renewed interest in January 2016. A study by Bradley Schaefer (Louisiana State University) of archived photographic plates, variously recorded between 1890 and 1989 and stored within the Harvard College Observatory, revealed that KIC 8462852 has been slowly dimming for at least the past 100 years [21]. This gradual decrease in brightness, by some 0.15 magnitude, is, again, completely unprecedented within the class of F spectral-type main sequence stars, and the result further exacerbates the problem of explaining KIC 8462852. Indeed, Schaefer argues that the slow, century-long dimming cannot be due to the same cometary break-up hypothesis that has been invoked as an explanation for the short-term brightness dips observed by the Kepler satellite.

Of course, it now seems inevitable that the story of KIC 8462852 cannot be so simply stated, and within 2 weeks of the analysis by Schaefer being posted, a second paper, by Michael Hippke and Daniel Angerhausen, appeared online [22]. This new analysis, using the same data archive as Schaefer, concluded that the century-long dimming of KIC 8462852 was a data artifact resulting from improper plate calibration. It need hardly be added that Schaefer immediately rejected the conclusions from the new analysis.

The most recent study, announced in August 2016, on the fading issue was by Benjamin Montet (Harvard-Smithsonian center for Astrophysics) and Joshua Simon (Carnegie Institution of Washington). These authors used actual Kepler satellite data to reveal that over the first 1000 days of observations, the measured energy flux from KIC 8462852 did, in fact, decrease at a rate of some 0.34 % per year, for a total decline of 0.9 %. From days 1000 to 1200 the measured flux decreased even more rapidly, dropping by more than 2 % over the time interval.

It is clear, therefore, that something long-lived and strange and unaccounted for is going on with respect to the brightness variations of KIC 8462852, but exactly how long-lived and how strange remains to be seen. Interestingly, Monet and Simon comment that, "No known or proposed stellar phenomena can fully explain all aspects of the observed light curve," but one can be sure that new and innovative ways of explaining the observations will be found as time goes by.

Located a mere 1500 light years away, KIC 8462852 provides us with another "WOW" SETI event. The observed light curve defies explanation in terms of standard and known astronomical phenomenon, and directly allows for a legitimate interpretation in terms of the expected signs of an advanced civilization capable of constructing huge structures. And, yet, other features of the star, its very young age and relatively large mass and luminosity compared to the Sun, argue against the possibility of any advanced civilization ever coming about. Perhaps one can circumvent the early origin of life on the assumed planet(s) in orbit about KIC 8462852 via the lithopanspermia hypothesis, but the rapid evolutionary rise to advanced intelligence (capable of building large space structures) within a timeframe of 1 billion years or less seems truly incredible. However, if humanity has learned just one thing about the universe and its marvelous contents, it is that it, and they, will never ceased to amaze.

The possible existence of extraterrestrial civilizations (intelligent or otherwise) within the Milky Way Galaxy remains an open question at the present time. The human species may be the only life form in the entire galaxy (universe?) capable of framing the question and having the technology to begin searching for potential interstellar neighbors. At 12 billion years old, however, our galaxy is still extremely young when compared to its expected life expectancy. The star formation rate within the Milky Way may well have peaked some 6 billion years ago, but stars will continue to form within its disk for the next 100 trillion years. We are presently just 0.006 % of the way through our galaxy's stellarific era. As time moves on, the great engines of cosmic recycling will continue to transmute and transform the interstellar medium, driving it towards an ever heavy-element richer composition, and this, in turn, will result in more and more planets being formed. Indeed, the future holds out the great promise for an ever-increasing abundance of planet-building material being available for the construction of life-supporting real estate.

Most of the future stars to form within the Milky Way will be of a low mass, having a K or M spectral type, and they will accordingly be long-lived. Combining the properties of stellar longevity with the metallicity-enhanced interstellar medium preference for producing more planets during the star formation phase the future prospects for intelligent life emerging somewhere within the galaxy look increasingly good. More time and more locations to work with will aid evolution in making those very hard and very low probability steps that take simple bacteria (panspermia supplied, or not) to intelligent entities capable of contemplating, searching, and theorizing about the heavens.

The prospects for testing many of our current underlying astrobiological assumptions and models took an important step forward in August 2016, with the announcement of a planet being detected in the habitability zone of our closest stellar neighbor—Proxima Centauri. The discovery of Proxima b, by Guillem Anglada-Escudé (Queen Mary University of London) along with an international assembly of some 30 coworkers, was announced in the journal *Nature*. The planet has an orbital period of 11.186 days, and if the planet has an atmosphere liquid water might exist on its surface.

The prospects for life having evolved and established an existence on Proxima b is a fundamental question ripe for future study, but it is intriguing to note that Proxima b is nearly a billion years older than Earth. The full picture, as of yet, remains unclear, but the system will, with little doubt, come under intense scrutiny in future years. We have now officially entered, however, the era in which the first alien landscape, atmosphere, and hopefully astrobiological signatures on another planet outside of the Solar System can be studied directly. Additionally, given the relative galactic closeness of Proxima to the Solar System it is possible that Proxima b might one day be imaged *in situ* by interstellar space probes. Indeed, just such an initiative, the Breakthrough Starshot Project (BSP), was announced in July of 2015. Financed by Yuri Milner and endorsed by Stephen Hawking, BSP plans to find ways of developing a "100 million miler per hour mission to the stars within a generation" (http://breakthroughinitives.org). The aim of the BSP is to develop both the power system and the instrumentation needed enable a mission to the α-Centauri system by the end of the twenty-first century. The vision is grand, the prize even grander, even grander, and let us hope that the apparently ever-shortening attention span of the human race stays the course.

The universe, as far as the story can be read at the present time, appears to be moving, though very, very slowly, towards a state of prescience in which the emergence of life and consciousness becomes ever more likely. Indeed, in a recent study published in the December 1, 2015, issue of the *Monthly Notices of the Royal Astronomical Society*, Peter Behroozi and Molly Peeples have used estimates of the planet formation rate within galaxies to argue that at the present epoch there are of order 10^{20} Earth-like planets, with about an equal number of giant, Jupiter-like planets within the observable universe. Using the classic Drake equation [23] as a guide to the possible number of civilizations that might exist at the present time, Behroozi and Peeples suggest that there is a less than 8 % chance we are the only intelligent civilization in the universe.

The narrative pertaining to life in the universe is long and difficult. There are no essay passages, words can be jumbled and strange, with intermittent twists of *Finnegan's Wake*, and so far humanity has hardly progressed beyond the prologue. The search for our intelligent neighbors will no doubt continue in one form or another, and humanity will presumably continue to stumble onwards towards an uncertain future—just as it has done for the past many millennia. Molecular clouds and star formation, the two pillars of creation, will continue to work their alchemy. Indeed, stars will continue to form in our galaxy for at least another 100 trillion years, and life, in all its understood, not-understood and presently unimagined forms, will continue to appear and evolve. Indeed, "We ain't seen nothin' yet."

You only live once, but if you do it right, once is enough.

—Mae West

REFERENCES

1. See J. W. Schopf and B. M. Packer's initial research paper on this topic: Early Archean (3.3-billion to 3.5 billion year old) microfossils from Warrawoona group, Australia. *Science*, **237**, 70-73 (1987). For a more recent analysis see, B. T. De Gregorio and T. G. Sharp's paper, Determining the biogenicity of microfossils in the Apex Chert, Western Australia, using transmission electron structures. *Lunar and Planetary Science XXXIV* (2003), pdf1267. Additionally, see the research paper by Herald Fumes and co-workers, Early life in Archean pillow lavas. *Science*, **304**, 578-581 (2004).
2. Elizabeth Bell *et al.*, Potentially biogenic carbon in a 4.1 billion-year-old zircon. Available at: www.pnas.org/cgi/10.1073/pnas.1517557112.
3. Alexei Sharov, Genome increase as a clock for the origin and evolution of life. *Biology Direct*, **1**:17, 2006. Additionally, see A. Sharov and R. Gordon's 2013 research paper, Life before Earth (arxiv.org/abs/1304.3381).
4. See the detailed research paper by Nora Noffke, Ancient sedimentary structures in the < 3.7 Ga Gillespie Lake, Mars, that resemble macroscopic morphology, spatial associations, and temporal succession in terrestrial microbialites. *Astrobiology*, **15**, 169-192 (2015).
5. See the highly speculative, but also highly thought-provoking book by Fred Hoyle and Chandra Wickramasinghe, *Astronomical Origins of Life: steps towards panspermia* (Kluwer Academic Publishing, 2000).
6. See the research paper by, B. Gladman *et al.*, Impact seeding and reseeding in the inner solar system, *Astrobiology*, **5**, 483 (2005).
7. See the research paper by W. Napier, A mechanism for interstellar panspermia, *Monthly Notices of the RAS*, **348**, 46 (2004).
8. See the research paper by E. Belbruno *et al.*, Chaotic exchange of solid material between planetary systems: implications for lithopanspermia (arxiv.org/pdf/1205.1059).
9. See the research paper by Karen Olsson-Francis and Charles Cockell, Experimental methods for studying microbial survival in extraterrestrial environments, *Journal of Microbiological Methods*, **80**, 1 (2010).
10. The original research paper is by David McKay *et al.*, Search for past life on Mars: possible relic biogenic activity in Martian meteorite ALH 84001, *Science*, **273**, 924-930

(1996). For a relatively recent up-date on the original ideas, see the article by Michael Schirber in the 21 October, 2010 issue of *Astrobiology Magazine* (http://www.astrobio.net/). Details of the Nakhla meteorite study are given in M. Fisk *et al.*, Iron-magnesium silicate bioweathering on Earth (and Mars), *Astrobiology*, **6**, 48–68 (2006).

11. See the research paper by Mauri Valtonen *et al.*, Natural transfer of viable microbes in space from planets in the extra-solar systems to a planet in our solar system and visa-versa, (2008): arxiv.org/pdf/0809.0378.

12. Another detection of a WOW-like signal was reported in August 2016 by astronomers working with the RATAN-600 radio telescope in Zelenchukskaya, Russia. The signal was actually detected in May of 2015 and lasted for 4 seconds. The researchers who discovered the signal argue that it was received from the direction to the Sun-like star HD 164595, which interestingly is known to harbor at least one Neptune-mass planet. Remarkably, given that the signal was from HD 164595 (at a distance of 95 light years), the recorded signal strength indicates that the transmission energy would amount to a staggering 10^{20} watts. Only a Kardashev Type I civilization could harness this amount of energy (comparable to one ten-thousandths that of the Sun's energy output). Follow-on observations of HD 164595 have been made since the initial detection, but no further *signals* have been recorded.

13. See Adam Frank and Woodrfuff Sullivan, Sustainability and the astrobiological perspective: Framing human futures in a planetary context. *Anthropocene* **5**, 32-41 (2014). See also the author's book, *Terraforming: the creating of habitable worlds* (Springer, 2009) – especially, Appendix D: population growth and lily word. Adam Stevens (The Open University) and co-workers have recently considered the possibility of detecting catastrophic global end-phases in the research paper, Observational signatures of self-destructive civilizations - arxiv.org/pdf/1507.08530.pdf. Nick Bostrom's papers can be found at http://nickbostrom.com.

14. See the research paper, Roger Griffith *et al.*, The Ĝ infrared search for extraterrestrial civilizations with large energy supplies III: the reddest extended sources of WISE, published in the *Astrophysical Journal Supplement* (**217**, article id. 25, 2015). Specifically Griffith and co-workers used the WISE catalog to search out so-called passive spiral galaxies. Such galaxies have low near-UV luminosity, look intrinsically red in the optical part of the spectrum, and have high mid-IR luminosity. The research paper by Garrett is, Application of the mid-IR radio correlation to the Ĝ sample and the search for advanced extraterrestrial civilizations, appearing in *Astronomy and Astrophysics*, **581**, L5 (2015).

15. See the research paper by R. A. Carrington, *IRAS*-based whole sky upper limit on Dyson spheres, published in *The Astrophysical Journal*, **696**, 2075 (2009). There is a large amount of literature concerning the possible structure of a Dyson sphere – the author has discussed such objects, for example, in the book *Rejuvenating the Sun and Avoiding Other Global Catastrophes* (Springer, New York, 2008).

16. Interstellar travel and the requirements for such distant adventures have been very ably discussed by K. F. Long in the book, *Deep Space Propulsion: a roadmap to interstellar flight* (Springer, New York, 2012). See also *Starship Century: toward the grandest horizon*, edited by James and Gregory Benford (Microwave Sciences, 2013). See also the research paper by Ulvi Yurtever and Steven Wilkinson, Limits and Signatures of Relativistic Spaceflight: arxiv.org/pdf/1503.05845v3.pdf.

17. Light curve generated with data and analysis programs provided by the NASA Exoplanet Archive website: http://archive.stsci.edu/kepler/data_search.php.

18. The initial research paper by T. S. Boyajian *et al.*, Planet Hunters X: KIC 8462852 – where's the flux? is published in the *Monthly Notices of the Royal Astronomical Society*, **457**, 3988-4004 (2016). See also, M. A. Thompson *et al.*, Constraints on the circumstellar dust around KIC 8462852, in *Monthly Notices of the Royal Astronomical Society*, **458**, L39 – L43 (2016). In addition, see the article by Eva Bodman and Alice Quillen, KIC 8462852: Transit of a large comet family: arxiv.org/pdf/1511.08821v2.pdf.

19. See the research paper by S. Sato *et al.*, Habitability around F-type stars: arxiv.org/pdf/1312.7431v2.pdf.

20. See the research article by G. R. Harp, *et al.*, Radio SETI observations of the anomalous star KIC 8462852: arxiv.org/pdf/1511.0.1606.pdf. In employing the survey results to dismiss the use of beamed microwave-driven spacecraft, the researchers really mean that no microwave beam pointed directly towards the ATA was recorded to the sensitivity limit of the receiving system. The no-detection signal does not actually rule out the use of such technology at KIC 8462852, and likewise the survey, which amounted to just a few hours of actual data recording, is not of sufficient length to truly rule out any alien (that is non-naturally modulated) signal.

21. See the research paper by Bradley Schaefer, KIC8462852 faded at an average rate of 0.165 ± 0.013 magnitudes per century from 1890 to 1989: arxiv.org/pdf/1601.03256v1.pdf. This paper, for which the entire scientific content is encapsulated within the title, is based upon an analysis of the in total 500,000 archival photographic plates contained in the Harvard College Observatory's collection. The brightness estimates are based upon 1232 plates that have been processed through the *Digital Access to a Sky Century at Harvard* (DASCH). Secondary brightness data was obtained via the visual analysis of 131 additional plates by Schaefer.

22. See Michael Hippke and Daniel Angerhausen paper, KIC 8462852 did likely not fade during the last 100 years, at: arxiv.org/pdf/1601.07314.pdf. The fading issue has additionally been taken-up by Benjamin Montet and Joshua Simon in an August 2016 research paper - arxiv:1608.01316v1.

23. Developed by radio astronomer and SETI pioneer Frank Drake in the early1960s, as a means of focusing attention towards the important developmental parameters, the Drake equation is used to estimate the number of active and communicative extraterrestrial civilizations in our galaxy. While various forms of the equation exist, the standard version asserts that the number of active civilizations within our galaxy N can be expressed as: $N = R^* \cdot fP \cdot NP \cdot fL \cdot fI \cdot fC \cdot L$, where R^* is the typical star formation rate, fP is the fraction of stars that have planets, NP is the average number of planets per star that can support life, fL is the fraction of planets that actually develop life, fI is the fraction of life bearing planets that produce intelligent life, fC is the fraction of intelligently habited planets that develop communication technologies that we might be able to detect, and L is the time over which active broadcasts are made. According to how one chooses the input numbers, the outcome of the Drake equation gives a value to N that falls anywhere between 1 (i.e. just us) to hundreds of millions. In many ways the actual value of $N \geq 1$ is irrelevant, and the real point of the equation is to emphasis the various channels, and the interconnectedness of those channels, that are likely to be of significance in the eventual emergence of a galactic civilizations. Provided the parameter choices do not result in a value of $N = 0$ at the present epoch (which would contradict our own existence) then the equation has effectively done its job.

Appendix

The Magnitude Scale and Interstellar Extinction

We live our lives and society is ordered through the adoption of arbitrary schemes. Certain colors, symbols, or words are used to denote some specific action or detail. The color red, for example, when seen on a traffic light indicates stop, but when seen on a tap it indicates hot water. If a word is written in red ink it means danger or beware; if a number is written in red ink it means you owe money.

The color choice is arbitrary. Green or purple could just as easily have been used as the impact color, but the point is, the color conveys non-verbal information, and it produces some specific action, and provided one knows what the convention is then the system makes sense and order prevails.

Astronomers use an arbitrary scheme for expressing the brightness of stars. The fact that the astronomers' brightness scheme is arbitrary, however, is decidedly odd. It is odd in the sense that it takes the fundamental quantity that is actually measured, the energy flux, and then converts it into another number based upon an arbitrary scale. Again, this is a strange thing to do, but it is a convenient action for several reasons. Most of all, however, since the mid-nineteenth century, it is the system that astronomers have adopted (for good or ill) and provided we stick with it, and understand where it came from, then order and consistency are maintained.

Hippacrchus of Nicaea compiled, circa 135 B. C., one of the first detailed star catalogs, inspired it is said, by the sudden appearance of a new star (supernova). The original catalog is now long lost, but its legacy endures—the legacy being the adoption of an arbitrary scheme for ranking the brightness, or magnitude, of the stars. Hipparchus chose six magnitude classes to describe the stars: magnitude 1 stars were the brightest, while magnitude 6 stars were the faintest visible to the human eye. The scheme is arbitrary in the sense that Hipparchus could have chosen ten, or five, or eight magnitude classes, but he chose six, and this was the basic star brightness division scheme (mostly) adopted by astronomers thereafter.

Jumping forwards in time by nearly 2000 years, astronomy was in the early days of revolution. Indeed, the late eighteenth century saw a fundamental change in the way astronomy was being done, moving further and further away from the eye as the basic measuring instrument and ushering in the introduction of non-human observing techniques such as photography and photometry. Perhaps not surprisingly, it was William Herschel who first attempted to use energy flux measurements to determine the relative

© Springer International Publishing AG 2017
M. Beech, *The Pillars of Creation*, Springer Praxis Books, DOI 10.1007/978-3-319-48775-5

brightness of stars. Indeed, the most fundamental characteristic of a star that is measurable from Earth is its energy flux, which is a measure of the amount of electromagnetic radiation received at the detector per unit area per unit time.

In the *SI* system the units of energy flux are joules/m²/s, and the larger the measured flux from a star the brighter it will appear. The fundamental shift in using flux measurements, however, is that it is based on a comparative instrument observation, rather than an observer's eye estimate (which is, of course, based on the observer's individual experience and visual acuity). For this comparative system to work consistently, however, at least one standard star is required, and everyone must agree to use this standard star in their studies.

One of the most important consequences of Herschel's flux measurement experiments was that the flux determined for a 1st magnitude star was found to be about 100 times larger than that received from a 6th magnitude star. It was this result that inspired Oxford University astronomer Norman Pogson, in an 1856 publication, to propose the magnitude scheme that is still in use to this very day. Pogson was interested in measuring brightness changes in asteroids, and was looking for a consistent way to gauge the observed brightness changes with respect to the phase angle, the angle subtended at the asteroid between Earth and the Sun. Pogson picked up on Herschel's flux results but imposed on it a strict variation. Pogson argued that if the flux measurements from two stars differ by a factor of 100, then those two stars must differ in brightness by exactly 5 magnitudes.

Additionally, and most importantly for the scheme to work, Pogson also introduced a set of standard comparative stars. These stars are taken as having a known and non-variable magnitude, and they are used as one of the stars in any comparative analysis. In this way the observer measures the flux from the star of interest, then the flux from any one of the standard stars, and the end result is the magnitude of the star of interest.

As a grand Victorian scientist, Pogson not only sought to improve the consistency of new brightness measurements. He also strove to link it to the ancient, classical Greek past. Pogson realized that Herschel's experimental results, his new scheme, and the brightness classes of Hipparchus could all be linked together by the mathematical transformation: $m = -2.5 \log(f) + C$, where f is the measured flux, m is the so-called apparent magnitude, and C is a constant. The logarithm term in the expression was based on the idea, which is not actually true, that the eye responds to brightness changes in a logarithmic fashion, while the 2.5 term is entirely a mathematical convenience to bring the scheme into line with the Hipparchus magnitude scheme.

The magnitude scheme outlined by Pogson works as follows. If we have two stars, one star being a standard star of magnitude m_S, then by measuring the flux from the star of interest f^* and the flux from the standard star f_S, the magnitude of the star of interest m^* is given by the relation: $m^* - m_S = -2.5 \log(f^*/f_S)$. We now see that if the two fluxes differ by a factor of 100 (i.e., $f_S = 100\,f^*$), so $m^* - m_S = -2.5 \log(1/100) = 5$, that is, the two stars are exactly five magnitudes different in brightness. Furthermore, if the standard star is a first magnitude star, where $m_S = 1$, the star of interest, in the case where the fluxes differ by a factor of 100, must be of magnitude $m^* = 1 + 5 = 6$, and accordingly, a flux ratio of 100 corresponds to a magnitude difference of 5.

If, in contrast, we take two stars, one of which has a measured flux twice that of the other (i.e., $f_1 = 2f_2$), then their magnitude difference will equal $m_1 - m_2 = -2.5 \log(2) = -0.75$. That is, star 1 is 0.75 magnitudes brighter than star 2. Alternatively, if two stars differ by a factor of 1 in their magnitudes, $m_1 - m_2 = 1$, then the star 2 will be 2.512 times brighter than star 1 (i.e., $f_2 = 2.512 f_1$).

Standardization follows from Pogson's scheme by adopting the arbitrary mathematical scheme and a set of standard stars of constant apparent magnitudes. With these two conditions in place, any telescope and flux measuring instrument at any observatory can be calibrated to produce consistent results. The choice of the standard star is entirely arbitrary, but of course, as with all arbitrary schemes some choices are better than others.

Clearly the Sun, while close and well studied, is not a good standard, mostly because it is far too bright (it blinds the eye) and, of course, it is not visible at nighttime. The North Star (Polaris = α *Ursae Minoris*) was originally taken as a standard magnitude +2 star, but Ejnar Hertzsprung found it to be of variable brightness in 1911. Likewise, Vega (α *Lyrae*) was initially adopted as a standard magnitude 0 star, but since observations with the Infrared Astronomy Satellite (IRAS) in 1983 revealed that it had a distinct infrared excess due to an orbiting dust disk, it, too, has been dropped as a standard star. All this goes to highlight the difficulty of establishing standard stars for instrument and observation calibration. A commonly used set of 526 standard stars distributed along the celestial equator is described by Arlo Landolt (Louisiana State University) in the July 1992 issue of the *Astronomical Journal*.

Most usefully, the apparent magnitude scale, as introduced by Pogson, can be extended beyond the classical range of 1–6. Indeed, the magnitude scale is such that the brighter an object is the larger and more negative is its apparent magnitude. The Sun has an apparent magnitude of −26.74, while the full Moon has an apparent magnitude of −12.9; the brightest star in the sky is Sirius (α *Canis Majoris*), and it has an apparent magnitude of −1.47. At the other end of the scale, the fainter an object is the larger and more positive is its apparent magnitude. The human eye can see as faint as about magnitude +6; the classic limit, while the faintest object observable with the Hubble Space Telescope, is +31.5.

The apparent magnitude of star is just a number, and it doesn't actually tell us anything useful about a star. It simply separates out the bright stars from the faint stars as they are seen from Earth. An observer somewhere else in the galaxy will deduce an entirely different set of magnitudes. This relates to the inverse distance squared law dependency of the flux. The luminosity L of a star is a fundamental measure of its energy output, being equal to the amount of energy radiated into space across the entire electromagnetic spectrum per unit time.

In the *SI* system the units for the luminosity are joules per second. When we measure the energy flux f, however, at a distance d from a specific star, the energy that the star has radiated into space each second is spread over the entire surface area of a sphere of radius d, and accordingly: $f = L/4\pi d^2$. As ever, there are caveats to the applicability of the formula just presented. It is assumed that the star radiates energy into space across the entire electromagnetic spectrum, from γ-rays to radio waves, that the flux is additionally measured over the same wavelength domain, and that the star radiates energy into space equally in all directions. It also assumes that no energy is absorbed or scattered out of the line of sight as the starlight travels between the star and the detector, and that the detector is perfectly efficient. For all this, however, with the flux being expressed in terms luminosity and distance, the magnitude scale can be converted into something more useful than just a brightness scale.

The absolute magnitude M is defined as being the apparent magnitude of a star at a distance of 10 pc. This distance is again arbitrary; astronomers could have chosen 1 or 100, or 23, but 10 is easy enough to work with. With this definition in place a relationship between the absolute magnitude of a star, its apparent magnitude, and distance can be derived, and we have: $m - M = 5\log(d/10 \text{ pc})$, where the units for distance measure (determined by parallax methods) are explicitly stated. The absolute magnitude of a star is directly related to its luminosity, and it is a quantity that can be calibrated against other observable stellar characteristics such as spectral type and luminosity class.

It is generally taken as standard that the apparent and absolute magnitudes are based on flux measurements made in the visual (V magnitude) part of the electromagnetic spectrum. The luminosity by definition, however, is based on the flux over all wavelengths of the electromagnetic spectrum, from gamma-rays to radio wavelengths. The luminosity of a star is determined, therefore, by applying a model-derived, temperature-dependent, bolometric correction BC term to M_V, with $M_{\text{Bol}} = M_V + BC$, and $\text{Log}(L/L_\odot) = 0.4 \, (4.79 - M_{\text{Bol}})$, where the 4.79 term is the calibrated absolute (bolometric) magnitude of the Sun.

Various photometry systems have been developed around Pogson's flux-magnitude scheme, and astronomers now characteristically use various sets of transmission filters to measure the energy flux at very specific wavelengths of light. One of the most common systems in use is that of the Johnson and Morgan UBV filter system introduced in 1953. In this system three transmission filters are used, one in the ultraviolet, one in the blue, and one in the visual part of the electromagnetic spectrum, and the flux from a particular star is measured through each of the filters in turn. From these various filter measurements color indices can be produced, such as $(B - V)$, which is the magnitude difference deduced from the flux measured through the B filter compared to the flux measured through the V filter. The $(B - V)$ color index is particularly useful in that it can be calibrated against temperature.

The Johnson and Morgan system has been extended into the infrared part of the spectrum with filters designated as I, J, K, and L. Although astronomers have long agreed to use the basic magnitude system developed by Pogson, they have been far from conservative with respect to its refinement concerning transmission filter systems. The Asiago Database on Photometric Systems (ADPS, maintained by the Osservatorio Astronomico di Padova, Italy) presently lists 218 different photometric systems that have been employed by astronomers since 1940.

The fundamental properties of main sequence stars are summarized in Table A.1, while Fig. A.1 shows a schematic HR diagram (Compare this to Figs. 5.2 and 5.3).

To account for the effects of interstellar dust the flux f_0 radiated by a star into space is modified by an exponential term dependent on the optical depth τ such that $f = f_0 e^{-\tau}$. With this correction term, the greater the optical depth the smaller the physically measured flux f compared to the true flux $f_0 = L/4\pi \, d^2$. The optical depth is further cast in terms of the opacity α, which is a measure of how the interstellar medium restricts the passage of radiation through it per unit meter length, and the distance d to the star: $\tau = \alpha \, d$.

Proceeding, as before, the difference between the apparent and absolute magnitude now becomes $m - M = 5 \log(d/10 \text{ pc}) + A \, (d/1000 \text{ pc})$, where $A = 2.5 \, \alpha \log(e) \, 1000$ pc is the extinction due to interstellar dust in magnitudes per kiloparsec. When the opacity of the medium is zero, corresponding to $\alpha = 0$, this indicates that there is no reduction in the measured flux compared to the true flux by the intervening space between the star and observer,

Table A.1   Characteristic properties of main sequence stars

Spectral type	$M/M_\odot$	$L/L_\odot$	$T(K)$	M_V	$(B-V)$	BC
O5	40	500,000	42,000	−5.7	−0.33	−4.40
B0	18	20,000	30,000	−4.0	−0.30	−3.16
B5	6.5	800	15,200	−1.2	−0.17	−1.46
A0	3.2	80	9790	+0.65	−0.02	−0.30
A5	2.1	20	8180	+1.95	+0.15	−0.15
F0	1.7	6	7300	+2.7	+0.30	−0.09
F5	1.3	2.5	6650	+3.5	+0.44	−0.14
G0	1.10	1.26	5940	+4.4	+0.58	−0.18
G2	1.00	1.00	5790	+4.7	+0.63	−0.20
G5	0.93	0.79	5560	+5.1	+0.68	−0.21
K0	0.78	0.40	5150	+5.9	+0.81	−0.31
K5	0.69	0.16	4410	+7.35	+1.15	−0.72
M0	0.47	0.063	3840	+8.8	+1.40	−1.38
M5	0.21	0.0079	3170	+12.3	+1.64	−2.73

The physical properties of mass and luminosity are in solar units and are taken from the *Handbook of Space Astronomy and Astrophysics* (Ed., M. V. Zombeck. CUP, 1990). The temperature and photometric data is from *Allen's Astrophysical Quantities* (Ed., A. N. Cox, Springer, 1999)

so $A=0$. As the opacity of the intervening medium increases, however, the value of A increases. Effectively, the extinction-corrected magnitude-distance formula tells us that the true apparent magnitude (i.e., if there were no extinction) is given by the relationship $m_{true} = m_{observed} - A$ $(d/1000\ \mathrm{pc})$. The value of A is generally determined empirically by star counting methods.

In order to determine the number of dust grains Ng in a specific line of sight, imagine that the interstellar dust grains (as shown in Fig. 3.18) have a radius of $a=3\times10^{-7}$ m (this is a characteristic size, some grains will be smaller, others larger). The effect of all these particles, seen in projection on the sky, will be to directly intercept or block from view Ng πa^2 of the starlight per unit area. A more exact expression would be written as $Ng\ Q(\lambda)\ \pi$ a^2, where $Q(\lambda)=Q(\mathrm{scattering})+Q(\mathrm{absorption})$ is a wavelength dependent function that takes into account the detailed scattering and absorption probabilities. In our analysis we assume $Q(\lambda)=1$ at all times. The amount of light (or energy flux) that gets through to our telescope will then accordingly be $(1-Ng\ \pi\ a^2)$. Expressed in terms of magnitudes, the dust extinction Δm will therefore be $\Delta m = -2.5\ \mathrm{Log}\ (1-Ng\ \pi\ a^2)$.

Now, in Chap. 1 we established that the number count of stars increases by a factor of about three each time the magnitude limit is increased by one (recall Table 1.1). Specifically, this result can be written in the form of a straight line relationship, such that $\mathrm{Log}(N)=\mathrm{con}$-stant $+0.48$ (m), where N is the number of stars counted to a limiting magnitude m and where the 0.48 is from the identity $3\approx10^{0.48}$. Therefore, if we plot a graph of the log of the number of stars counted up to a specified magnitude, we should find a straight line relationship of slope 0.48. Furthermore, if we perform star counts in equal area regions of the sky, one in a field direction where there is no (or at least very little) dust absorption, and one in the direction of a dust cloud, the resultant number versus magnitude plot will look like Fig. A.2, where line (a) corresponds to the field star counts and line (b) corresponds to the star counts in the direction of the nebula.

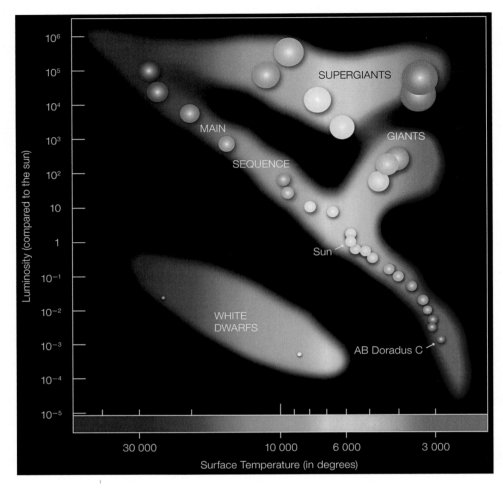

Fig. A.1 A schematic HR diagram showing the relationship between temperature and luminosity for main sequence stars. The high temperature stars are also the most luminous, while the low temperature stars are the least luminous. Small-sized stars are located towards the *lower lefthand corner* of the diagram, while the largest stars plot towards the *upper righthand corner*. The large and luminous giant and supergiant stars plot above the main sequence, while the white dwarfs, which are on a cooling sequence, plot below it. The star AB Doradus C is a T Tauri star and forms part of a pre-main sequence quadruple-star system. Image courtesy of ESO

Importantly, the slopes of the two lines shown in Fig. A.2 will be the same and equal to 0.48. Additionally we see that the slope of each line corresponds to the ratio $\Delta \log(N)/\Delta m$, and this can be rearranged to reveal that $\Delta m = 2.1 \, \mathrm{Log}(N_0/N)$, where N is the star count in the direction of the nebula and N_0 is the star count in the dust-free region of the field. The factor of 2.1 is the inverse of 0.48. Combining our two expressions for Δm now enables us to make an estimate of the total number of grains in the line of sight. Specifically, we have $Ng = [1 - N_0/N)^{-0.84}]/\pi \, a^2$.

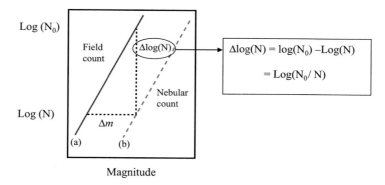

Fig. A.2 Schematic star number count versus magnitude diagram for two regions, one with no extinction (**a**) and the other (**b**) with an extinction amounting to Δm magnitudes

In the case of the Coalsack Nebula considered in Chap. 3 (recall Fig. 3.19) we found $N=210$ and $N_0=1800$, which indicates that $\Delta m \approx 2$, and that $Ng \approx 3 \times 10^{12}$ grains per square meter in the sky. Furthermore, given that the Coalsack Nebula is some 600 light years distant from the Sun, and that the circular region used to determine the star count N has a diameter of 5 minute of arc (or 1/12 of a degree), the nebula has a physical cross-sectional area of some 6×10^{31} m^2 of the sky. Combining this area estimate with Ng, the number of grains per unit area indicates a total grain count of 1.8×10^{44} grains in the nebula. Furthermore, given that each grain has a mass of order 2.5×10^{-16} kg, the total mass of dust grains will be about 4.5×10^{28} kg, which in turn is equivalent to 1/22 the mass of the Sun. And finally, given that the dust mass in a typical nebula is estimated to be about 1/10 that of the mass of the gas in the nebula, the Coalsack Nebula region studied in Fig. 3.19 has an estimated total mass of some 0.25 solar masses.

Index